Vacuum Physics and Techniques

Physics

PHYSICS AND ITS APPLICATIONS

Series Editor

E. R. Dobbs
University of London

S. B. Palmer
Warwick University

This series of short texts on advanced topics for students, scientists and engineers will appeal to readers seeking to broaden their knowledge of the physics underlying modern technology.

Each text provides a concise review of the fundamental physics and current developments in the area, with references to treatises and the primary literature to facilitate further study. Additionally texts providing a core course in physics are included to form a ready reference collection.

The rapid pace of technological change today is based on the most recent scientific advances. This series is, therefore, particularly suitable for those engaged in research and development, who frequently require a rapid summary of another topic in physics or a new application of physical principles in their work. Many of the texts will also be suitable for final year undergraduate and postgraduate courses.

1. Electrons in Metals and Semiconductors
   R. G. Chambers

2. Basic Digital Electronics
   J. A. Strong

3. AC and DC Network Theory
   A. J. Pointon and H. M. Howarth

4. Nuclear and Particle Physics
   R. J. Blin-Stoyle

5. Thermal Physics
   Second edition
   C. B. P.Finn

· 6. Vacuum Physics and Techniques
   T. A. Delchar

7. Basic Electromagnetism
   E. R. Dobbs

Vacuum Physics and Techniques

T. A. Delchar

Lecturer
Department of Physics
University of Warwick
UK

CHAPMAN & HALL

London · Glasgow · New York · Tokyo · Melbourne · Madras

Published by Chapman & Hall, 2–6 Boundary Row, London SE1 8HN

Chapman & Hall, 2–6 Boundary Row, London SE1 8HN, UK

Blackie Academic & Professional, Wester Cleddens Road, Bishopbriggs, Glasgow G64 2NZ, UK

Chapman & Hall Inc., 29 West 35th Street, New York NY10001, USA

Chapman & Hall Japan, Thomson Publishing Japan, Hirakawacho Nemoto Building, 6F, 1–7–11 Hirakawa-cho, Chiyoda-ku, Tokyo 102, Japan

Chapman & Hall Australia, Thomas Nelson Australia, 102 Dodds Street, South Melbourne, Victoria 3205, Australia

Chapman & Hall India, R. Seshadri, 32 Second Main Road, CIT East, Madras 600 035, India

First edition 1993

© 1993 T. A. Delchar

Typeset in 10/12 Times by Expo Holdings, Malaysia
Printed in Great Britain by St Edmundsbury Press, Suffolk

0 412 46590 6

A catalogue record for this book is available from the British Library

Library of Congress Cataloging-in-Publication data
Delchar, T. A.
    Vacuum physics and technology / T. A. Delchar. — 1st. ed.
        p. cm. — (Physics and its applications; 6)
    Includes index.
    ISBN 0-412-46590-6 (alk. paper)
    1. Vacuum. 2. Vacuum technology. I. Title. II. Series.
QC166.D45   1993
533'.5—dc20   92-21136
                                                               CIP

♾ Printed on permanent acid-free text paper, manufactured in accordance with the proposed ANSI/NISO Z 39.48–199X and ANSI Z 39.48–1984

Contents

Preface   ix

1 Some aspects of kinetic theory   **1**

1.1   Introduction   1
1.2   Maxwell's distribution of velocities   2
1.3   Ideal gas laws   2
1.4   Non-equilibrium properties of gases   4
1.5   The mean free path   5
1.6   Viscosity of gases at ordinary pressures   7
1.7   Viscosity of gases at low pressures   9
1.8   Slip coefficient   9
1.9   Viscosity of gases at very low pressures   11
1.10 Thermal conductivity at ordinary pressures   13
1.11 Thermal conductivity at low pressures   14
1.12 Temperature jump distance   14
1.13 Accommodation coefficient   16
1.14 Free-molecule heat conduction   19
1.15 Diffusion of gases   21
1.16 Molecule collision frequency   23
1.17 Thermal transpiration (thermomolecular flow)   23
      Problems   24

2 Flow of gases through tubes and orifices   **26**

2.1   Introduction; the Knudsen number   26
2.2   Flow conductance and impedance   28
2.3   Viscous flow   29
2.4   Molecular flow   32
2.5   Effusion and molecular flow through short tubes;
      the Clausing coefficient   33
2.6   Flow in the transition range; Knudsen flow   38
2.7   Free-molecule conductance of tubes in series   42

2.8   Flow in vacuum systems; the speed of a pump   44
      Problems   45

3   Physisorption, chemisorption and other surface effects   47

3.1   Introduction   47
3.2   Physisorption   48
3.3   Chemisorption   56
3.4   The condensation coefficient   59
3.5   The sticking probability   59
3.6   Getters and gettering   61
3.7   Sputtering   63
3.8   Sorbents and molecular sieves   65
3.9   Electrical clean-up   69
3.10 Electron and ion stimulated desorption   72
3.11 Diffusion   73
3.12 Permeation   74
3.13 Outgassing   76
      Problems   77

4   Vacuum pumps; the physical principles   78

4.1   Introduction   78
4.2   Types of mechanical pumps   80
4.3   The rotary oil pump   80
4.4   Hook and claws pump   86
4.5   Roots-type pump   87
4.6   The molecular drag pump, turbomolecular pumps   88
4.7   Diffusion pumps   94
4.8   Getter pumps and getter-ion pumps   100
4.9   Sorption pumps   104
4.10 The cryopump   105
      Problems   108

5   Pressure measurement   110

5.1   Introduction; the vacuum spectrum   110
5.2   Absolute gauges   112
5.3   Spinning rotor gas friction gauge   120
5.4   Pirani gauge   122
5.5   Thermocouple gauge   125

5.6	Thermistor gauge	126
5.7	Ionization gauges	127
5.8	The hot cathode ionization gauge	130
5.9	The Bayard–Alpert gauge	131
5.10	Modulated Bayard–Alpert gauge	133
5.11	The high pressure ion gauge	135
5.12	The Penning gauge, cold cathode ionization gauge	136
5.13	Magnetron ionization gauges	137
5.14	Hot cathode magnetron gauge	139
5.15	Calibration for different gases	142
5.16	Partial pressure gauges and residual gas analysers	143
5.17	Magnetic deflection analysers	144
5.18	The omegatron	146
5.19	Monopoles, quadrupoles and ion traps	148
	Problems	152
6	**Vacuum system design**	**153**
6.1	Review of vacuum system components	153
6.2	Gas flow between chamber and pump	154
6.3	Residual gas sources	156
6.4	Ultimate pressure, speed of exhaust	157
6.5	Pump-down time	158
6.6	Design of a high vacuum system	161
6.7	Backing pumps	163
6.8	High vacuum pumps	166
6.9	Ultrahigh vacuum	167
	Problems	174
7	**Construction accessories and materials**	**175**
7.1	Introduction	175
7.2	Static pipe couplings and seals	175
7.3	Dynamic vacuum seals	181
7.4	Indirect motion techniques	183
7.5	Vacuum valves	186
7.6	Valveless gas admission	189
7.7	Glass/ceramic–metal seals	190
7.8	Optical windows and electrical feedthroughs	192
7.9	Baffles, cold traps and sorption traps	194
7.10	Metals for vacuum use	196

8 Leak detection **200**

8.1 Introduction 200
8.2 Reviewing the symptoms 204
8.3 The leak detection routine 206
8.4 Leak detection using a test gas 208
8.5 Leak detection using partial pressure analysis 211
    Problems 211

9 Vacuum systems in practice **213**

9.1 Introduction 213
9.2 The portable leak detector 214
9.3 Vacuum-based coating systems 220
9.4 Molecular beam production 225
9.5 A molecular beam system 229

Appendix A Standard vacuum technology symbols **231**

Appendix B Brief outline of some vacuum material properties **235**

Appendix C Vacuum equipment manufacturers and suppliers **243**

Answers to problems **245**

Further reading **246**

Index **248**

Preface

Vacua are fundamental to a range of scientific explorations and techno-logical processes; the range is now very wide indeed, extending from studies of the properties of atomically clean surfaces at pressures which verge on extreme ultrahigh vacuum, $< 10^{-11}$ mbar, to freeze-drying of foodstuffs at pressures which may be only $\sim 10^{-1}$ mbar. In the pressure range between, vacua are required for TV tube production, vacuum impregnation, vacuum furnaces, vacuum coating, storage ring opera-tion, semiconductor processing, etc. There have been tremendous advances and improvements made in vacuum technology, particularly in the last 20 years. It is the purpose of this book to try to introduce the reader to the basic physics which underlies modern vacuum tech-nology, and then show how it is applied in vacuum production and measurement. As far as possible everything described in this book is in current use (there are a few exceptions which are included because they illustrate some interesting physics).

The author is only too aware of the omissions which have been made to keep this book to a comfortable size and which vacuum experts will doubtless spot. Hopefully, despite these omissions, the reader will end up feeling able to contemplate the design and/or specification of a simple vacuum system, with some confidence that the design will meet requirements and some understanding of the physical (and chemical) processes which are operating.

Finally, the author would like to thank the two people without whose help the preparation of this book would have been impossible, namely, Melanie Henley, who typed (and retyped) the manuscript, and Jeanette Chattaway, who prepared the drawings. Thanks are also due to the University of Warwick for the grant of Study Leave covering part of the time involved in the preparation of this book.

T. A. Delchar, 1992

1

Some aspects of kinetic theory

1.1 INTRODUCTION

For a proper understanding of gaseous processes, especially at low pressures, it is essential to consider them using the kinetic theory of gases. This theory assumes that gases are made up from atoms or molecules which, for a given material, are all alike with regard to size, shape and mass. In addition, these atoms/molecules are in constant motion, travelling in rectilinear paths the directions of which are only changed by collisions, either with other molecules or with the walls of their enclosure. These latter collisions must, on average, be perfectly elastic otherwise there would be a gradual change in the kinetic energy of the molecules in the enclosure. An important consequence of this collisional motion is that the molecules may exert pressure, or transport energy, momentum or mass. Any analysis of these phenomena has, as a prerequisite, a knowledge of the distribution of velocities. Here we must turn to Maxwell's (1860) analysis. It is not appropriate within the remit of this volume to indulge in a formal derivation of the results of the Maxwell distribution of velocities and these results will merely be quoted and used. The interested reader will find complete derivations in one of the standard texts, for example, Guénault (1988). Under the conditions encountered in vacuum systems, all gases may be considered as ideal or perfect gases and Maxwell–Boltzmann statistics always apply.

1.2 MAXWELL'S DISTRIBUTION OF VELOCITIES

Maxwell's distribution law enables us to define three distinct speeds, namely, the most probable speed

$$c_{max} = \left(\frac{2k_B T}{m}\right)^{1/2} \tag{1.1}$$

the average speed

$$\bar{c} = \left(\frac{8k_B T}{\pi m}\right)^{1/2} \tag{1.2}$$

and the root mean square speed

$$c_{rms} = \left(\frac{3k_B T}{m}\right)^{1/2} = \sqrt{\bar{c^2}} \tag{1.3}$$

In each equation k_B is the Boltzmann constant, 1.38×10^{-23} JK^{-1}, T the absolute temperature and m the mass of the atom or molecule.

1.3 IDEAL GAS LAWS

If a molecule of mass m and speed c collides with a surface, it is assumed to rebound with the same speed and the consequent momentum change is $2mc$. If the number of molecules striking unit area in unit time with average speed c is v, then the total impulse on unit area is $2mcv$ and must equal the pressure exerted thus

$$2mcv = p \tag{1.4}$$

Suppose we have n molecules per unit volume, then at any instant of time we may assume that the molecules are moving in one of the six directions corresponding to the six faces of a cube. Therefore, on average, $\frac{1}{6}nc$ molecules will cross unit area in unit time, and (1.4) may be rewritten as

$$p = \frac{1}{3}mnc_{rms}^2 \tag{1.5}$$

If we denote the product mn by ρ, the density, we can write

$$p = \frac{1}{3}\rho c_{rms}^2 \tag{1.6}$$

which tells us that, at constant T, pressure varies directly as the density, or inversely as the volume. Now the total kinetic energy of the molecules in a volume V is

$$\frac{1}{2} mnc_{rms}^2 V = \frac{3}{2} pV \tag{1.7}$$

The average kinetic energy of the molecules must be the same for all gases at any given temperature. This allows us to define temperature in terms of the average kinetic energy of the molecules, thus

$$\frac{1}{2} mc_{rms}^2 = \frac{3}{2} k_B T \tag{1.8}$$

Combining (1.5) and (1.8) yields

$$p = nk_B T \tag{1.9}$$

which is an extremely useful relation since it tells us that equal volumes of all gases at any given values of temperature and pressure contain an equal number of molecules; a fact that was first pointed out by Avogadro in 1811. Equation (1.9) in fact summarizes all of the experimentally determined gas laws, so that Boyle's law is obtained by multiplying both sides by V and noting that nV is the total number of molecules. Alternatively, Charles' law is obtained if we now maintain p and the number of molecules constant.

Avogadro's law may be rewritten to express the fact that the molar mass M of any gas is that mass in g, calculated for an ideal gas, which occupies, at 273 K and one atm pressure, a volume of 22.414 dm^3 (the molar volume). Thus for hydrogen, M is 2.016 g, or for helium 4.003 g. We can therefore rewrite the equation of state for an ideal gas as

$$pV = \frac{M_g}{M} RT \tag{1.10}$$

where M_g is the mass in g, M is the molar mass in g and R is the universal gas constant, 8.31 Jmol^{-1}K^{-1}. Here also, V is the volume of M_g grams of gas at pressure p and temperature T. Equation (1.10) may be conveniently re-expressed as

$$pV = n_M RT \tag{1.11}$$

where n_M is the number of moles in the volume V at a given value of p and T.

For the molar volume V_0 we can combine (1.9) and (1.11) to give

$$pV_0 = N_A k_B T \qquad (1.12)$$

where N_A is Avogadro's number and $N_A = R/k_B$. Similarly, we note that the mass per molecule, m, is given by

$$m = M/N_A$$

where N_A has the value 6.022×10^{23} mol^{-1} and m is, of course, in grams.

In the MKS or SI system of units, the unit of pressure is the newton per square metre, or pascal (Pa). In the cgs system of units, the unit of pressure is the dyne cm^{-2} or microbar. The standard atmosphere is defined as the pressure exerted by a column of mercury 760 mm high, that is to say that 1 atm = 760 mm Hg. This standard is often expressed in torr rather than mm, where 1 torr is essentially equal to 1 mm Hg, to within 1 part in 7 million (1 standard atmosphere is defined as 101 325 kNm^{-2} exactly and 1 torr is just 1/760 of a standard atmosphere).

The relationship between the various systems of pressure measurement may be demonstrated quite easily, as follows:

$$1 \text{ newton} = 10^5 \text{ dyne and } 1 \text{ m}^2 = 10^4 \text{ cm}^2$$

therefore

$$1 \text{ Nm}^{-2} = 10 \text{ dyne cm}^{-2} = 10 \text{ } \mu\text{bar}$$

1 mm Hg $= 0.1 \times 13.5951 \times 980.665$ dyne cm^{-2} = 1333.22 microbar = 133.32 Nm^{-2}.

If in (1.9) n is expressed in units of m^{-3}, k_B in joules per kelvin and T in kelvins, then p will be given in units of pascals (Pa).

These four different units of pressure measurement, the bar, torr, mm Hg and pascal, are all in widespread use. In recent scientific literature the pascal is usually, but not invariably, used. In manufacturers' specifications for equipment the mbar is the most commonly used unit and the one adopted here for the description of pumps and vacuum gauges. Conversion from one set of units to another is straightforward since **1 mbar = 100 Pa.**

1.4 NON-EQUILIBRIUM PROPERTIES OF GASES

It has already been pointed out that an adequate understanding of kinetic theory is central to any appreciation of vacuum physics. Of particular importance in this connection are the non-equilibrium properties of gases, namely, the transport of energy (thermal conductivity), the transport of momentum (viscosity) and the transport of mass (diffusion). It turns out that there are strong similarities between these transport properties and that they may be employed, for example, as the basis of pressure measuring systems and are central to the understanding of the mode of operation of some pumps.

In order to analyse and discuss transport properties we shall need the relationships derived from Maxwell's (1860) analysis of the distribution of velocities and, in addition, some idea of the mean free path of gas atoms or molecules under any given set of conditions. Thus, the atom/molecule mean free path, λ, will usually be the index by which we shall define the likely behaviour of our vacuum system.

1.5 THE MEAN FREE PATH λ

We can readily obtain some feeling for the mean free path λ by imagining one molecule threading its way through a cluster of stationary molecules; molecules are assumed to be smooth, hard, elastic spheres. If the diameter of the moving molecule is σ then it cannot approach nearer than 2σ to the centre of any adjacent molecule without undergoing a collision as shown in Fig. 1.1(a). Put in another way, we may consider an exclusion disc of diameter 2σ centred upon the moving molecule, which sweeps out a cylindrical exclusion volume $\pi \bar{c} \sigma^2$ in unit time, Fig. 1.1(b). The number of collisions per unit time will simply be $n\pi\sigma^2 \bar{c}$, where n is the number of molecules per unit volume. The mean free path λ will then be given by the average distance between collisions, thus:

$$\lambda = \frac{1}{\pi n \sigma^2} \qquad (1.13)$$

An alternative view, which is readily adapted to dealing with mixtures of gases, is the following. The target area presented to a molecule by another of the same size is $\pi\sigma^2$; this is the exclusion disc noted previously. Now if there are n molecules per unit volume, then a slab of area A and thickness dx contains just $nAdx$ molecules each with target area $\pi\sigma^2$, making the effective target area $nAdx(\pi\sigma^2)$. When the effective target area equals A there is bound to be a collision. Let us suppose that this happens when $dx = \lambda$, then

$$nA\lambda\pi\sigma^2 = A$$

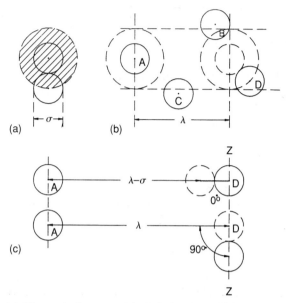

Fig. 1.1 (a) The exclusion zone (shaded) for the collision of two identical molecules, diameter σ. (b) The collision free path traced out by molecule A between molecules B and C before colliding with molecule D after distance λ, the mean free path. Molecules B, C and D are assumed stationary. (c) The variation in λ as position of target molecule D moves along line ZZ. The difference in λ is the diameter of target molecule D as the angle between the centres varies from 0° to 90°.

yielding the previous result. Again we are assuming that all molecules except one are stationary and may be considered hard spheres. The mean free path evaluated above is that corresponding to a very fast molecule which completes its path in a time so short that no molecules have entered or left the exclusion volume; this is the stationary molecule assumption. The true value of λ can be calculated if the distribution function of the velocities of the molecules is known. It obviously can never exceed $1/n\pi\sigma^2$ and indeed is not a well-defined quantity since, as reference to Fig. 1.1(c) shows, it varies by σ as the angle between the line of centres at the time of collision varies from 0° to 90°.

If we wish for a better value for λ we must take into account the motion of molecules. Thus if the molecules have a Maxwell distribution, this reduces the mean free path by a factor of $1/\sqrt{2}$, so that

$$\lambda = \frac{1}{\sqrt{2}\pi n\sigma^2} \tag{1.14}$$

If all the molecules have the same velocity, then the multiplying factor becomes $\frac{3}{4}$ and we have

$$\lambda = \frac{3}{4\pi n\sigma^2} \tag{1.15}$$

Equation (1.14) is due to Maxwell and (1.15) is due to Clausius (1857). It is useful to note that for air at room temperature, λ is conveniently derived from the relationship

$$\lambda = \frac{6.6}{p} \tag{1.16}$$

where p is in pascals and λ in mm.

1.6 VISCOSITY OF GASES AT ORDINARY PRESSURES

In this context, ordinary indicates that molecule–molecule collisions are dominant and λ is very much less than the basic dimensions under consideration. Experimentally, we find that for steady flow of gas down a tube the pressure drop is balanced by friction at the walls and the speed of the gas is governed by frictional or viscous forces between gas layers in relative motion. In practice, of course, gas layers mix due to random motion of molecules, i.e. molecules from the slow stream mix with those from the fast stream, tending to equalize speeds. We define the coefficient of viscosity η by means of the equation

$$\frac{F}{A} = \eta \frac{du}{dy} \tag{1.17}$$

where F is the shear force operating on area A in the gas and du/dy is the resulting velocity gradient. In Fig. 1.2(a) we consider two parallel plates of area A, separated by a distance h. Plate A_1 is assumed at rest while plate A_2 moves to the right with velocity u under the influence of the shear force F, which is required to overcome the viscous drag. We should expect that F will be proportional to A. The force per unit area, F/A, on either plate is $\eta\, du/dy$ from the definition of η. The sense of the force is in the direction of u on the moving plate and, of course, in the opposite direction for the fixed plate. The fixed plate is thus gaining

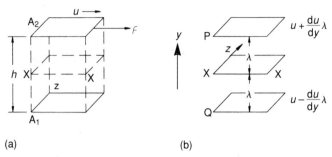

Fig. 1.2 (a) The disposition of plates A_1 and A_2 to produce a viscous flow at ordinary pressures; A_2 is moving at velocity u as a consequence of the shearing force f. (b) Velocity distribution between test planes P and Q in a viscous flow with velocity gradient du/dy.

momentum from the gas and the moving plate losing momentum to the gas at the rate $\eta\, du/dy$ per unit area. The viscous process can thus be described as a flow of momentum across the gas, a typical transport process. In order to calculate the coefficient of viscosity we must make three further assumptions.

1. Flow velocity is very small compared with the mean gas speed $\bar{c}$.
2. One sixth of the molecules cross area A.
3. The only molecules reaching plane XX in Fig. 1.2(b) are those which last had a collision a distance λ away.

The number of molecules crossing area A in unit time must therefore be

$$\frac{n\bar{c}A}{6}$$

From Q these molecules bring flow momentum to the plane XX of

$$m\frac{n\bar{c}A}{6}\left(u - \frac{du}{dy}\lambda\right)$$

From P, by similar reasoning, we get flow momentum

$$m\frac{n\bar{c}A}{6}\left(u + \frac{du}{dy}\lambda\right)$$

The net transport of momentum per second through the plane XX is thus

$$\frac{1}{3} mn\bar{c} A \lambda \frac{du}{dy}$$

which must equal the force F. Hence

$$F = \frac{mn\bar{c}}{3} \lambda A \frac{du}{dy} \tag{1.18}$$

and comparing with (1.17) we can write that

$$\eta = \frac{1}{3} mn\bar{c} \lambda \tag{1.19}$$

If we measure the gas density in units of m^{-3}, the molecular mass in kg, the velocity in ms^{-1} and the mean free path in m, η will have units of Pa s. It is worth noting from this result that viscosity is proportional to mean speed and hence $\sqrt{T}$, but is independent of pressure; it has dimensions $ML^{-1}T^{-1}$. This prediction that the viscosity of a gas will increase with increasing temperature is striking, bearing in mind that the opposite behaviour is found for liquids. At low pressures, however, the viscosity is not independent of pressure, as we shall see.

1.7 VISCOSITY OF GASES AT LOW PRESSURES

It is sufficient for the moment to define low pressures as those for which the mean free path becomes comparable with the distance between moving surfaces. There are three situations we can consider which provide limiting behaviour. In the first, already considered, we have effectively ignored the moving surfaces A_1, A_2, but as only half the transport process is operating at the walls, there must be some change in flow rate if the stress is to be transmitted to the wall. In the second case we suppose that the impinging molecules acquire a tangential velocity equal to that of the wall. In the third we assume no transfer of momentum to the surfaces.

1.8 SLIP COEFFICIENT

Although the velocity gradient is linear in the bulk of the gas, this is not true close to the plates. Hence the drift velocity of the gas near to the stationary boundary does not decrease linearly to zero, but varies in some way rather as shown in Fig. 1.3. The increased velocity near the boundary increases the momentum brought up to the surface and also

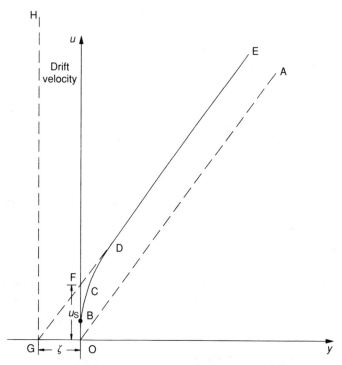

Fig. 1.3 The slip distance ζ for viscous flow over a stationary plate with assumed velocity distribution EF, true velocity distribution BCDE, slip velocity OF. Velocity zero assumed to occur at O. Assumed velocity at surface is u_S. In effect the stationary surface is considered displaced to GH.

the tangential force acting upon it. This increase is necessary in order to give a consistent result, since it is clear that if the velocity gradient is constant and linear right up to the surface, there will be a drag of $\frac{1}{2}F$ instead of F. Generally, the disturbance due to the wall will not extend beyond a few mean free paths and the correction for 'slip' is negligible, except at low pressures.

In order to gain some feel for the 'slip' effect it is convenient to make the assumption that the velocity gradient du/dy is everywhere the same and that the wall is displaced to the left in Fig. 1.3 by a distance ζ. This is equivalent to saying that the drift velocity near the wall can be approximated by the simple expression

$$u = u_S + \left(\frac{du}{dy}\right) y \qquad (1.20)$$

in which u_S is the slip velocity at the real wall and u is the drift velocity at a distance y from the wall. The value of u_S must be chosen to give the correct force on the wall. The rate at which the molecules of the gas are bringing up tangential momentum to each unit area of the walls is $\frac{1}{2}\eta du/dy$ due to the velocity gradient, and mu_S for each of the $\frac{1}{4}n\bar{c}$ molecules which are striking the wall in unit time, due to slip. If we equate these to the known drag $\eta du/dy$, we find that our expression for F/A has to be replaced by

$$\frac{F}{A} = \eta \frac{u}{(y+2\zeta)} \quad (1.21)$$

where $\zeta = \frac{2}{3}\lambda$. The viscosity will now appear to have decreased.

We can express the situation more completely by considering the fraction θ of molecules adsorbed and re-emitted with the tangential velocity of the wall. If this is done then we can write ζ more explicitly as

$$\zeta = 2C\lambda \frac{(2-\theta)}{\theta} \quad (1.22)$$

where C is a constant, close to 0.5.

It is apparent from (1.22) that ζ is always of the order of the mean free path λ, hence ζ must vary inversely with pressure, behaviour which was originally confirmed by Kundt and Warburg in 1875. The value of θ, which is in effect the transfer ratio for momentum, or diffuse reflection coefficient, will depend on the nature of the interaction between the gas molecules and the surface.

For a surface which reflects molecules specularly, i.e. with no change in tangential velocity, $\theta = 0$ and $\zeta = \infty$, viscosity cannot obtain a grip upon the wall and we have zero velocity gradient. Alternatively, the situation where molecules are reflected randomly with complete loss of their initial tangential velocities corresponds to $\theta = 1$, and ζ is almost equal to λ. Physically therefore, ζ is really the ratio of the internal friction of the gas (i.e. the coefficient of viscosity η) to the coefficient of external friction against the walls. The latter term, and hence ζ, must depend on the nature of the interaction with the wall.

1.9 VISCOSITY OF GASES AT VERY LOW PRESSURES

The concept of 'slip' as discussed above is only applicable when the layer of gas is many mean free paths thick so that ordinary viscous motion can come into existence. When this condition is not satisfied, we find that the treatment of flow is rather complex unless we use the free-molecule case wherein the density is low enough, or the gas layer is thin enough, that molecule–molecule collisions may be neglected. In this situation, we can write that the free-molecule viscosity of the gas, z say, is given by

$$z = p\left(\frac{m}{2\pi k_B T}\right)^{1/2} \frac{\theta}{(2-\theta)} \tag{1.23}$$

and

$$F/A = zu \tag{1.24}$$

Readers requiring a detailed explanation of this result should see Kennard (1938). Note, z has dimensions $ML^{-2}T^{-1}$. The momentum transferred per unit area per unit time is just zu, and molecules make the transition from one surface to the other without intermediate collisions, indeed, the free-molecule viscosity is independent of the spacing between plates or the apparent velocity gradient.

An interesting aspect of this result is that at very low pressures the rate of transference of momentum from a moving surface to another adjacent, parallel surface is directly proportional to pressure p and to the velocity u of the moving surface, a result originally demonstrated by Langmuir in his molecular pressure gauge, and exploited nowadays in the spinning rotor viscosity gauge; the reader is referred to section 5.3.

To summarize this discussion of momentum transport in gases, we can define three pressure regimes according to the value of the mean free path λ compared with the distance h, say, between the moving surfaces. If $\lambda \ll h$, then the coefficient of viscosity η has the dimensions $ML^{-1}T^{-1}$ and viscosity is independent of pressure but increases with temperature (contrast the behaviour of liquids). If $\lambda \leq h$ then the concept of slip appears, defined by the slip coefficient ζ, which has the dimensions of length. Now the viscosity is reduced in magnitude. The slip coefficient is of the order of the mean free path and must, therefore, vary with pressure, as does the viscosity. The nature of the moving surfaces is now important also. It is convenient to consider slip as equivalent to increasing the separation of the moving surfaces by an amount equal to the sum of the slip coefficients at each surface. If $\lambda > h$ we have the free-molecule situation where viscosity is proportional to pressure, the velocity of the moving surface and its nature. Viscosity now decreases with increasing temperature. The free-molecule viscosity, z, has dimensions $ML^{-2}T^{-1}$.

In the next chapter these transport regimes will be redefined in terms of the Knudsen number K_n.

1.10 THERMAL CONDUCTIVITY AT ORDINARY PRESSURES

We define thermal conductivity K according to the equation

$$\frac{dQ}{dt} = KA\frac{dT}{dy} \qquad (1.25)$$

Here Q is the net heat transport, A is the cross sectional area through which heat is being transported and dT/dy is the temperature gradient driving the process. An expression for the thermal conductivity of gases at ordinary pressures may be obtained by considering the energy transport between two test planes R and S in the gas separated by a distance 2λ and between which a temperature gradient dT/dy exists, as shown in Fig. 1.4.

If c_V is the specific heat at constant volume, per unit mass, then the transport of heat from R to S is just

$$\frac{n}{6}\bar{c}Amc_V\left(T + \frac{dT}{dy}\lambda\right)$$

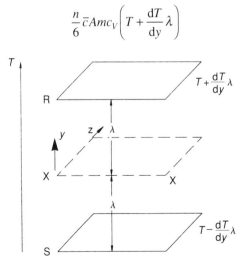

Fig. 1.4 The temperature distribution between test planes R and S at ordinary pressures, where $\lambda \ll h$, the plate spacing.

while the transport of heat from Q to P is similarly

$$\frac{n}{6}\bar{c}Amc_V\left(T - \frac{dT}{dy}\lambda\right)$$

The net transport of heat is then given by

$$\frac{dQ}{dt} = \frac{n}{3}\bar{c}Amc_V\lambda\frac{dT}{dy} \tag{1.26}$$

Comparing (1.25) and (1.26) we can write that

$$K = \frac{n}{3}\bar{c}mc_V\lambda \tag{1.27}$$

Note that K is independent of pressure and moreover simply related to viscosity by

$$K = \eta c_V \tag{1.28}$$

When η has the units of Pa s, and c_V has units $Jkg^{-1}K^{-1}$, then K will have units of $Wm^{-1}K^{-1}$.

1.11 THERMAL CONDUCTIVITY AT LOW PRESSURES

We have seen that both viscosity and thermal conductivity are independent of pressure in those situations where $\lambda \ll h$, i.e. high pressures. The parallel between the behaviour of the two properties exists also at low pressures where they both become pressure dependent. We shall see also that the concept of 'slip', introduced for the treatment of viscous processes at low pressures, has a thermal analogue, the so-called 'temperature jump distance', while θ (the diffuse reflection coefficient), has a thermal analogue, the accommodation coefficient.

1.12 TEMPERATURE JUMP DISTANCE

In our discussion of viscosity (momentum transport) the notion of 'slip' was introduced, and molecules were considered to slip with velocity u_S relative to the containing surfaces when the flow velocity gradient normal to the surface was du/dy. In effect we said that

$$u_S = \zeta \frac{du}{dy} \tag{1.29}$$

where ζ is the coefficient of slip.

By analogy with the phenomena of viscous slip we might expect to find a corresponding temperature discontinuity at low pressures, which we can define in a similar manner to yield a coefficient g known as the temperature jump distance, thus

$$T_S - T_1 = g \frac{dT}{dy} \tag{1.30}$$

This effect is displayed in Fig. 1.5 and was first demonstrated experimentally by Smoluchowski in 1898, although the existence of this

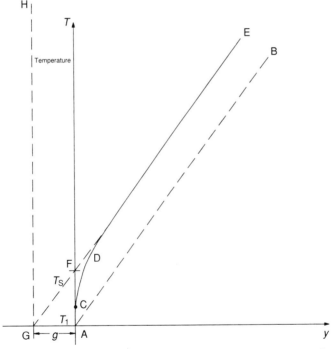

Fig. 1.5 The temperature jump distance g for a surface at temp T_1. Line EF is the assumed temperature distribution and CDE is the actual temperature distribution; assumed temperature at the surface is T_S. The surface at temperature T_1 is assumed displaced to GH.

effect had originally been suggested by Poisson. Smoluchowski developed a theory for the temperature jump, which introduced a constant that measured the extent to which interchange of energy takes place when a molecule of the gas strikes a solid surface. Knudsen, in 1911, also looked at the temperature jump problem and he introduced a slightly different constant, the accommodation coefficient, which is now widely used.

1.13 ACCOMMODATION COEFFICIENT

In considering the problems of heat transfer in gases at low pressure, it is often desirable to make use of a quantity which represents the extent to which energy transfer takes place when a gas molecule strikes a surface. This quantity, known as the accommodation coefficient, is the one originally introduced by Knudsen and is defined as

$$\alpha = \frac{T_0 - T_2}{T_0 - T_1} \tag{1.31}$$

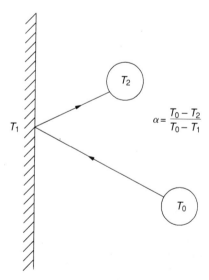

Fig. 1.6 The definition of the accommodation coefficient α for molecules colliding with a surface at temperature T_1. Here, T_0 is the temperature of the incoming molecule and T_2 is the temperature of the reflected molecule.

where T_0 is the temperature of the incident molecule, T_2 the temperature of the reflected molecule and T_1 the temperature of the surface upon which the molecule impinges, as shown in Fig. 1.6. If the molecules reach thermal equilibrium with the surface before escaping, that is $T_2 = T_1$, then $\alpha = 1$. Alternatively, if the molecules are reflected elastically without undergoing any change in energy, $T_2 = T_0$ and $\alpha = 0$. The temperature T_2 is not necessarily well-defined unless the scattered molecules have a Maxwellian distribution. Indeed, of course, a collection of molecules without a Maxwellian distribution does not have a temperature. As a consequence it is often convenient to replace the temperature appearing in (1.31) by the mean energies of the different species. Thus we obtain

$$\alpha = \frac{E_0 - E_2}{E_0 - E_1} \tag{1.32}$$

where E_0 is the mean energy of the approaching molecules and E_2 that of the reflected molecules, while E_1 is the mean energy of the molecules if they reach equilibrium with the surface before escaping.

In (1.32) $E_0 - E_2$ represents the net energy delivered to the surface by the gas molecules. This is also the net heat transfer across a plane in the gas, so that if we consider unit surface area and unit time we have

$$E_0 - E_2 = K \frac{dT}{dy} \tag{1.33}$$

where K is the thermal conductivity of the gas and dT/dy is the temperature gradient.

The quantity $E_0 - E_1$ is the difference between the incident energy and the energy removed by the molecules, assuming that they reach thermal equilibrium with the surface at temperature T_1. We can isolate two terms in this quantity, namely,

1. the excess energy carried by the molecules because of the temperature gradient dT/dy; this term is $\frac{1}{2}K dT/dy$;
2. the difference in energy content of two gas streams at temperatures T_S and T_1. The translational energy of the molecules crossing unit area in unit time in a gas at temperature T is $2k_B Tp/(2\pi m k_B T)^{1/2}$ where $p/(2\pi m k_B T)^{1/2}$ is just the collision frequency per unit area, per unit time; see equation (1.49).

The previous expression can be written more compactly as $2RTG$, where G is the mass of gas crossing unit area in unit time and equals $p/(2\pi RT)^{1/2}$, and R equals $\boldsymbol{R}/\boldsymbol{M}$ or the gas constant for one gram of gas.

If U is the internal energy of the molecules in unit mass of gas at a temperature T, then the total energy flow E_T across unit area in unit time is $G(2RT+U)$. In order to obtain the difference in energy flow for two different temperatures, T_S and T_1 (Fig. 1.5) we can write this as

$$G[2R(T_S-T_1) + (U_S-U_1)]$$

and for small temperature differences U_S-U_1 becomes

$$\frac{dU}{dT}(T_S - T_1)$$

so that

$$E_T = G\left[c_V + \frac{R}{2}\right](T_S - T_1) \qquad (1.34)$$

where we have substituted for $2R + dU/dT$ using the relationship for the specific heat at constant volume c_V given by

$$c_V = \frac{3}{2}R + \frac{dU}{dt} \qquad (1.35)$$

Now using (1.32) and (1.33) plus some manipulations we can write

$$K\frac{dt}{dy} = \alpha\left[\frac{K}{2}\frac{dT}{dy} + \frac{p}{(2\pi RT)^{1/2}}\left(c_V + \frac{R}{2}\right)(T_S - T_1)\right] \qquad (1.36)$$

Rearranging and using $c_V(\gamma - 1) = R$, where γ is the ratio of the principal specific heats of the gas, we have

$$T_S - T_1 = \left[\left(\frac{2-\alpha}{\alpha}\right)\frac{(2\pi RT)^{1/2}}{p}\frac{K}{(\gamma+1)c_V}\right]\frac{dT}{dy} \qquad (1.37)$$

The variable T in the above equation can be taken as the wall temperature provided the difference between T_S and T_1 is not too large. Comparison with (1.30) gives

$$g = \frac{(2-\alpha)}{\alpha}\frac{(2\pi RT)^{1/2}}{p}\frac{K}{(\gamma+1)c_V} \qquad (1.38)$$

In fact g can be rewritten using (1.27) to give

$$g = C'\lambda\frac{(2-\alpha)}{\alpha} \qquad (1.39)$$

where C' is a constant lying in the range 1.5–1.9, depending on the gas chosen. The similarity between (1.22) and (1.39) is striking. If $\alpha = 0$, i.e. $T_0 = T_2$ in Fig. 1.6, then $g = \infty$, whilst if $\alpha = 1$, $g \simeq \lambda$. Contrast the behaviour of ζ as θ varies from 0 to 1. The thermal conductivity now depends on pressure through g, so that if we consider two parallel plates set a distance h apart and at temperatures T_1 and T_2 respectively, we can write that

$$\frac{1}{A}\frac{\mathrm{d}Q}{\mathrm{d}T} = \frac{K(T_2 - T_1)}{h + g_1 + g_2} \tag{1.40}$$

where g_1, g_2 are the temperature jump distances for the plates 1, 2, as illustrated in Fig. 1.7. The terms g_1, g_2 vary with λ and hence inversely with pressure p. As a consequence, as p decreases, $\mathrm{d}Q/\mathrm{d}T$ will decrease below its high pressure value, T_1, T_2, and h assumed constant.

1.14 FREE-MOLECULE HEAT CONDUCTION

It is convenient to consider free-molecule heat conduction for the case of two parallel plates, Fig. 1.8. Since we have free-molecule conduction, molecules traverse the gap between the plates R and S without

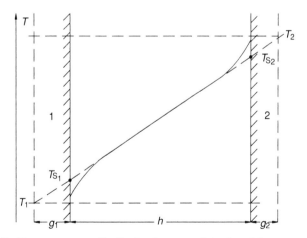

Fig. 1.7 The temperature distribution between plates 1 and 2 at temperatures T_1, T_2, having jump distances g_1 and g_2; T_{S_1} and T_{S_2} are the assumed temperatures at surfaces 1 and 2 respectively. The effective separation distance between the plates becomes $h + g_1 + g_2$ (at pressures where $h \geq \lambda$) thus lowering the temperature gradient. The values of g_1 and g_2 vary with λ and hence pressure.

undergoing collisions. We assume accommodation coefficients α_R, α_S respectively, and that the molecules may be divided into two groups, one group travelling from R→S, the other in the direction S→R. Each group is assumed Maxwellian corresponding to temperatures T_0 and T_1.

If we calculate the energy content of the two streams using the approach leading to (1.34), we can also show that the net energy transfer, per unit area per unit time, from R to S is

$$\frac{1}{A}\frac{dQ}{dT} = G\left(c_V + \frac{R}{2}\right)(T_1 - T_0) \qquad (1.41)$$

where the symbols have their previous meaning; dQ/dT now depends on pressure directly through G, and will depend on T_1 and T_0 through the accommodation coefficients for the two surfaces. The heat transfer is now independent of the plate separation.

We may summarize the behaviour of energy transport (thermal conductivity) with pressure in much the same way as we did for viscosity. Thus, at high pressure, $h \gg \lambda$, K is independent of pressure but increases with $\sqrt{T}$. When $h \geq \lambda$ the idea of a temperature jump distance appears. Now K depends on p and is reduced below its expected value by an apparent increase in separation of the heated surfaces, i.e. the temperature gradient is, in effect, reduced. When $h < \lambda$, K depends on pressure but becomes independent of the temperature gradient. Note the marked similarities with momentum transport.

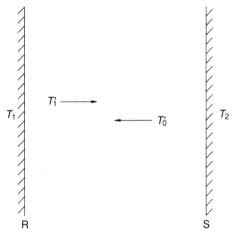

Fig. 1.8 The free-molecule flow situation for surfaces R and S at temperatures T_1 and T_0. Groups of molecules transit without collisions at temperatures T'_1 and T'_0.

The dependance of K on p affords a simple means for pressure measurement exploited in the Pirani gauge (section 5.4), while the energy transport between surfaces at different temperatures leads to the molecular radiometer force which is the basis of the Knudsen gauge, an absolute pressure gauge (section 5.2).

1.15 DIFFUSION OF GASES

When a gas contains two or more different kinds of molecules whose relative densities vary from point to point, a process known as diffusion is seen to occur whereby the composition of the gas becomes uniform. Alternatively, if we have a gas where different parts of it are at different densities, the density non-uniformities will be removed in time by diffusion, or mass transport.

Consider the situation where the density variation exists in the y direction only. We find that the net number N of molecules transported through a plane of area A in unit time is proportional to both the area A and the concentration gradient in the y direction, dn/dy. Thus:

$$N \propto A \frac{dn}{dy} \tag{1.42}$$

If the proportionality is removed we can write

$$N = -DA \frac{dn}{dy} \tag{1.43}$$

where D is the diffusion coefficient and the minus sign implies that flow is down the concentration gradient. From Fig. 1.9, the number of molecules moving from plane Q to plane P is just

$$A \frac{\bar{c}}{6} \left(n + \frac{dn}{dy} \lambda \right)$$

where n is the number of molecules/unit volume. Similarly, the number moving from P to Q is

$$A \frac{\bar{c}}{6} \left(n - \frac{dn}{dy} \lambda \right)$$

Net transport is just the difference between these two figures,

$$\left[-A \frac{\bar{c}}{3} \lambda \frac{dn}{dy} \right]$$

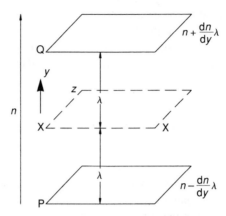

Fig. 1.9 The density distribution between test planes P and Q set 2λ apart in a density gradient of dn/dy.

so that we can write

$$-A\frac{\bar{c}}{3}\lambda\frac{dn}{dy} = -DA\frac{dn}{dy} \tag{1.44}$$

Hence,

$$D = \frac{\bar{c}}{3}\lambda \tag{1.45}$$

Note that D depends on p^{-1} as well as $T^{3/2}$ and $m^{-1/2}$.

If we wish to describe the diffusion coefficient appropriate to two gases mixing then the expression must be modified to yield

$$D = \frac{1}{3}\left(\bar{c}_1\lambda_1\frac{n_2}{n} + \bar{c}_2\lambda_2\frac{n_1}{n}\right) \tag{1.46}$$

where $n = n_1 + n_2$, and n_1, n_2 are the concentrations/unit volume of the species 1 and 2 respectively, λ_1, λ_2 are the associated mean free paths and $\bar{c}_1$, $\bar{c}_2$ are the mean speeds. This formula, originally derived by Meyer in 1899, predicts substantial variation of the diffusion coefficient with gas composition.

Finally, we note that at ordinary pressures we can write the viscosity coefficient η as $\frac{1}{3}\rho\bar{c}\lambda$ where ρ is the gas density. Thus

$$D = \eta/\rho \tag{1.47}$$

1.16 MOLECULE COLLISION FREQUENCY

An important result, originally derived by Meyer, shows that the number of gas molecules which strike unit area in unit time is given by

$$v = \frac{1}{4} n \bar{c} \qquad (1.48)$$

Substituting for n and $\bar{c}$ yields

$$v = \frac{p}{\left(2\pi m k_B T\right)^{1/2}} \qquad (1.49)$$

where p is the gas pressure. This is otherwise known as the Hertz–Knudsen equation. At a pressure of 1 Pa the collision frequency (molecules $cm^{-2}s^{-1}$) is $2.63 \times 10^{20}/(MT)^{1/2}$. Alternatively we can write that the mass of gas G incident on unit area in unit time is

$$G = mv = \frac{1}{4} \rho \bar{c} = \rho \left(\frac{8 k_B T}{\pi m} \right)^{1/2} \qquad (1.50)$$

1.17 THERMAL TRANSPIRATION (THERMOMOLECULAR FLOW)

Consider the situation depicted in Fig. 1.10 where two chambers A and B are at different temperatures T_A, T_B and separated by a small orifice or tube. If the pressures are sufficiently high that the mean free path λ is small compared with the tube radius a, then $p_A = p_B$, while the gas densities ρ_A and ρ_B are related by

$$\frac{\rho_A}{\rho_B} = \frac{T_B}{T_A} \qquad (1.51)$$

This result arises from the fact that when $\lambda \ll a$, a transition region many free paths thick must exist between the regions where the temperature has steady values T_A, T_B. There is no sharp temperature discontinuity in the connecting tube or orifice.

The situation is quite different, however, in the low pressure region where $\lambda \gg a$. The temperature difference over a free path is now significant and collisions among molecules passing through the orifice are unlikely. Under these conditions we can assume a temperature discontinuity at the orifice plane and, if we equate the gas flows in each direction, we obtain from (1.48)

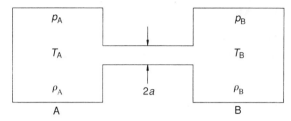

Fig. 1.10 Containers A and B connected by a tube of diameter $2a$. Pressure, density and temperature in each container are p_A, ρ_A and T_A; p_B, ρ_B and T_B respectively.

$$\frac{1}{4}n_A\bar{c}_A = \frac{1}{4}n_B\bar{c}_B$$

and since $p = nk_BT$ and $\bar{c} = (8k_BT/\pi m)^{1/2}$ we get

$$\frac{p_A}{p_B} = \left(\frac{T_A}{T_B}\right)^{1/2} \tag{1.52}$$

which is an apparent contradiction of the result of (1.51).

If, in Fig. 1.10, we increase the length of the connecting tube whilst still maintaining the condition $\lambda \gg a$, then gas will flow from the region of lower to the region of higher temperature, but at equilibrium, $p/T^{1/2}$ is constant along the tube and (1.52) describes the overall situation. If the system is not closed, then gas flow will continue indefinitely, provided the temperature gradient is maintained.

The transport of gas from a low to a high temperature zone is known as thermal transpiration and it must be taken into account whenever attempts are made to measure pressure in vacuum systems where the vacuum gauge is remote from the region in which the pressure must be determined and where a temperature differential exists between the two.

PROBLEMS – CHAPTER 1

1.1 An atomically clean titanium surface, at 300 K, is exposed to oxygen at a pressure of 10^{-5} Pa for 10 s. What fraction of the titanium surface will be covered with adsorbed oxygen *atoms* if every molecule which collides with the surface sticks? How long would it take for a clean titanium surface to become completely covered with oxygen atoms if the pressure were reduced to 10^{-8} Pa? Note: you should assume one atom of oxygen per titanium atom and 10^{15} titanium atoms cm^{-2}.

1.2 An isolated nitrogen-filled reaction chamber is immersed in liquid oxygen at 90 K and evacuated to an indicated pressure of 0.1 Pa. If the pressure gauge is at room temperature (300 K) and is connected to the reaction chamber by a connecting tube of 10 mm diameter, then calculate the true pressure in the reaction chamber. The molecular diameter of a nitrogen molecule is 0.37 nm.

1.3 A thin, flat plate, area 10^{-2} m, is suspended on the end of a fine wire spring (spring constant 2.0 Nm^{-1}) so that it is parallel with, and close to, an identical fixed plate. Both plates are initially at 300 K and sealed within a cell evacuated to 10^{-2} Pa. If the fixed plate is heated to 400 K, what change will there be in the extension of the spring? What change in extension will occur if the pressure is doubled?

1.4 A sealed tube contains helium which exerts a pressure of 10^{-5} Pa at liquid nitrogen temperature (77 K). By how much will the gas mean free path change if the pressure is increased by a factor of four, through raising the tube temperature to 308 K?

2

Flow of gases through tubes and orifices

2.1 INTRODUCTION: THE KNUDSEN NUMBER

The description of gas flow in vacuum systems is customarily categorized by a dimensionless number known as the Knudsen number, after its originator M. Knudsen. We have already seen in our examination of kinetic theory that the behaviour of the transport properties of a gas, such as thermal conductivity or viscosity, can vary with mean free path, λ, when λ becomes comparable with the separation of the surfaces involved. The Knudsen number introduces an explicit way to characterize gas properties as a function of pressure; we define the Knudsen number, K_n, as the ratio of the mean free path of a molecule to a characteristic dimension of the channel or pipe through which the gas is flowing.

For vacuum systems operating at high pressure, where the mean free path is small compared with the characteristic dimension of the pipe or channel (small Knudsen number), collisions between molecules are much more common than collisions between molecules and the pipe walls. The characteristics of the gas flow are then necessarily determined by the intermolecular collisions. Consequently, the gas can be considered to be a continuous medium, or viscous fluid, since the properties of the gas do not vary noticeably in one free path. Under these conditions flow may be analysed and described hydrodynamically; we have in effect viscous flow.

The converse situation exists at low pressures where the mean free path is large compared with the characteristic dimension (large Knudsen number). Here, the gas flow is limited by molecular collisions with the walls of the channel and we are faced with a situation where analysis of the flow reduces to an essentially geometric problem of determining the restrictive effects of the walls on the motion of a molecule. Intermolecular collisions are now more or less absent so that each molecule can be considered to act independently of all the others. Flow under conditions of large Knudsen number is thus known as free-molecular flow, or commonly just molecular flow.

Between the regions of viscous flow and molecular flow we can expect some sort of transitional behaviour which will exhibit the properties of both regions to some extent, since both types of collision are involved. Unfortunately, it is not possible to derive flow equations for this pressure range from first principles; we can only describe the flow by semi-empirical equations. This makes flow in the so-called transition range, Knudsen flow, more difficult to handle.

Hitherto, the Knudsen number has merely been specified as either 'large' or 'small'. Before we can use it, however, we must clearly be able to assign actual limiting values to the Knudsen number for each flow region. For a tube of circular section the characteristic dimension has historically been the radius, a (although some authors use the tube diameter), and the Knudsen number, K_n, is given by λ/a. In practice it turns out that we can consider the flow to be viscous when $\lambda/a < 0.01$, whilst the flow will be molecular when $\lambda/a > 1.0$. The transition range flow occurs when $0.01 < \lambda/a < 1.0$ as shown in Fig. 2.1, which defines the flow spectrum.

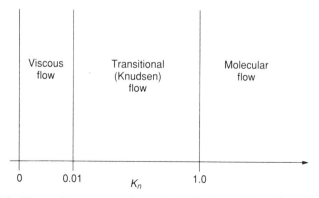

Fig. 2.1 The gas flow spectrum for a tube of circular section, radius a, defined in terms of the Knudsen number K_n, where $K_n = \lambda/a$, the mean free path divided by the tube radius.

The flow equations appropriate to each region are considered in the following sections, sometimes without derivation since such material does not fall easily within the range of this text. The standard textbooks which consider the basis of the flow equations in detail are noted under 'Further Reading'.

2.2 FLOW CONDUCTANCE AND IMPEDANCE

In any study of flow we will be interested in the flow rate, or throughput. We define the flow rate Q_F by the equation

$$Q_F = p \frac{dV}{dt} \tag{2.1}$$

where dV/dt is the volumetric flow rate and p the pressure at which it is measured. The units for Q_F will typically be Pa m^{3}s^{-1} using the SI system, or perhaps mbar ls^{-1} otherwise. Because 1 Pa = 1 Nm^{-2} and 1 J = 1 Nm, the units of Q_F may be more simply expressed as Js^{-1} or watts (1Pa m^3 s^{-1} = 1 W). Therefore we can consider throughput or flow rate as the energy per unit time crossing a plane in the system. It should be noted that the power flow is equivalent to the mass flow rate, but only if the system is everywhere at a constant known temperature; a spatial temperature change can alter the energy flow rate without altering the mass flow.

We can use the ideal gas law to substitute for p and V in (2.1) yielding

$$Q_F = k_B T \frac{dN}{dT} \tag{2.2}$$

where T is the absolute temperature and dN/dt is the rate at which molecules cross the plane in which we are measuring pressure. If we construct an electrical analogue, and we shall find this a most convenient approach, then Q_F corresponds to a molecular current.

We now define a quantity, the conductance, F, given by

$$F = \frac{Q_F}{p_2 - p_1} \tag{2.3}$$

Here p_2 is the upstream pressure at the entrance to the tube and p_1 is the downstream pressure at the exit from the tube. The tube is maintained at a uniform, fixed temperature. If we extend the electrical analogy then

p_2-p_1 is the potential difference driving the flow, and since we have already identified Q_F with the current, F must be analogous to conductance. It follows that $1/F$ is the resistance or impedance to flow for which we shall generally use the symbol Z. If two or more different tubes are connected in series we are tempted to use our electrical analogue to say that the total impedance will now be given by the sum of the impedances, so that

$$\frac{1}{F} = \frac{1}{F_1} + \frac{1}{F_2} \qquad (2.4)$$

Alternatively, for parallel connection we will have

$$F = F_1 + F_2 \qquad (2.5)$$

In practice though, (2.4) and (2.5) must be used with some care. This applies particularly to (2.4), as we shall see in section 2.7. The SI units of conductance are m^3s^{-1}, but it is not unusual to find conductance specified in ls^{-1}; 1 m^3s^{-1} is equivalent to 1000 ls^{-1}.

The practical application of the above concepts reduces to the need to determine appropriate expressions for F in the different flow regions. In each case, of course, F will have a particular representation for a given tube or channel structure and will reach its simplest form for circular section tubes.

2.3 VISCOUS FLOW

In the case of Knudsen numbers < 0.01 we can treat the flow by use of Poiseuille's equation. This equation involves four basic assumptions, namely

1. Flow is streamlined, i.e. not turbulent.
2. Flow is fully developed; the flow velocity profile is constant throughout the length, ℓ, of the tube. This last condition imposes restrictions on ℓ, which cannot be made small.
3. The flow velocity at the wall of the tube must be zero.
4. The gas stream is considered incompressible. The criterion for condition (4) is generally established by asking the question, 'What conditions will make compressibility important?'. It can be shown that compressibility is not important provided $\frac{1}{2}M_n^2 \ll 1$, where M_n is the Mach number of the flow velocity. The Mach number is defined as the ratio of the flow velocity v_F to the velocity of sound in the gas, v_S. As a practical limit we can take a value of $M_n = \frac{1}{3}$.

The average flow velocity across a plane in a circular tube of radius a, where the pressure is p, is defined by

$$v_F = \frac{Q_F}{\pi a^2 p} \tag{2.6}$$

where we have used (2.1), noting that dV/dt in this case is just $v_F \pi a^2$, and πa^2 is the cross sectional area of the tube. Since $v_F/v_S = M_n$ and $M_n \simeq \frac{1}{3}$ we can set an upper limit for Q_F given by

$$Q_F = \frac{1}{3} \pi a^2 p v_S \tag{2.7}$$

Poiseuille's equation can be derived rather simply for circular section tubes, merely from a consideration of the definition of viscosity in (1.17) as applied to the situation depicted in Fig. 2.2. Inserting the appropriate quantities we can establish dv/dr, the velocity gradient in this case, and integrate to find v, thus

$$v = \frac{(p_2 - p_1)\pi(a^2 - r^2)}{4\eta \ell} \tag{2.8}$$

We require the volumetric flow rate dV/dt which can be obtained by multiplying (2.8) by the cross sectional area of the annulus, $2\pi r dr$, and then integrating between 0 and a. This procedure yields

$$\frac{dV}{dt} = \frac{(p_2 - p_1)\pi^2 a^4}{8\eta \ell} \tag{2.9}$$

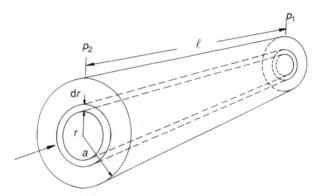

Fig. 2.2 Analysis of viscous flow in a circular section tube.

Returning to (2.1) and writing $p_a = \frac{1}{2}(p_2 - p_1)$ we obtain

$$Q_F = \frac{p_a\left(p_2 - p_1\right)\pi^2 a^4}{8\eta\ell} \qquad (2.10)$$

but from our definition of F, (2.3), we can write

$$F_V = \frac{p_a \pi a^4}{8\eta\ell} \qquad (2.11)$$

This result shows that the conductance, F_V, of a circular section pipe in the viscous flow region is dependent on the average pressure p_a, which may be difficult to define. This is particularly important if it is intended to join different size tubes end to end, since the abrupt change in diameter may well change the total pressure difference between the ends of the tube. In consequence, the formulae for determining flow conductance, (2.4) and (2.5), may not be accurate. Finally, it is important to have some idea of what is meant by a 'long' tube, assumption (2), in which the flow profile is fully developed and therefore constant. Clearly, when a gas flows into a tube from a large volume, to begin with the flow velocity is uniform over the whole of the tube mouth. This situation cannot be maintained in the presence of viscous drag since the flow velocity at the walls will become zero once the velocity profile is established and maintained. When the flow profile becomes constant it is said to be fully developed, but we need some guide to the conditions which must be met in order to reach this state; in particular we need to know the length of tube over which the entrance transition occurs.

Many authors have calculated the effect of the entrance transition, but a simple guide due to Langhaar (1942) is that the transition to fully developed flow will occur in a distance ℓ_T given by

$$\ell_T = 0.227 a R_n \qquad (2.12)$$

where R_n is the dimensionless Reynolds number of the flow. We can express R_n in terms of simple physical quantities already defined, so that

$$R_n = \frac{m}{\pi \eta k_B T} \frac{Q_F}{a} \qquad (2.13)$$

For a given gas ℓ_T will depend primarily on Q_F, and the effect of the transition to fully developed flow is to lower the flow rate for a given pressure difference.

A similar effect is found if Q_F is increased overmuch, since the Reynolds number then exceeds a critical value and the flow becomes turbulent. If the entrance to a tube is free of sharp corners, i.e. it is well-rounded and smooth, the critical value of R_n will be around 1000. If this critical value is exceeded it will be necessary to increase the tube radius, a, to restore viscous flow. Fortunately, the conditions for turbulent flow can only be met in a conventional vacuum system, in the backing or roughing line of a large pump during the initial pump down from atmospheric pressure; it is not otherwise important.

2.4 MOLECULAR FLOW

Under conditions where the mean free path is greater than the characteristic dimension, the rate of flow is determined by collisions of molecules with the tube walls rather than molecule–molecule collisions. Most of the original treatment of flow under these conditions is due to Knudsen (1909, 1911), with later improvements due to Smoluchowski.

Knudsen's approach to the problem of molecular flow was to balance the pressure at the ends of a tube against the momentum transferred to the walls. Thus, for a 'long' tube of length ℓ, varying cross section A, and perimeter H, Knudsen deduced a fundamental relation for the flow rate Q_F as

$$Q_F = \frac{4}{3} \frac{\bar{c}}{\int_0^\ell \frac{H}{A^2} \, \mathrm{d}\ell} (p_2 - p_1) \qquad (2.14)$$

It follows, using (2.3), that the molecular flow conductance, F_M, is given by

$$F_M = \frac{4}{3} \frac{\bar{c}}{\int_0^\ell \frac{H}{A^2} \, \mathrm{d}\ell} \qquad (2.15)$$

and is now independent of pressure; contrast viscous flow. This freedom from pressure dependence makes molecular flow conductance a much more useful concept than its equivalent in the viscous flow region. In this context, a 'long' tube is one in which the length of the tube is roughly 100 times the largest transverse dimension, as shown in Fig. 2.5.

It is usually a rather straightforward process to perform the integral in (2.15) corresponding to a particular tube configuration. Thus, for a circular tube of constant radius a, $H = 2\pi a$ and $A = \pi a^2$ so

$$F_M = \frac{2}{3}\frac{\bar{c}\pi a^3}{\ell} \tag{2.16}$$

A very convenient approximation for a cylindrical tube is that due to Yarwood, namely, $F_M = 100\, a^3/\ell$, ls^{-1}, if a and ℓ are in cm.

For a rectangular tube with sides a and b, we have

$$H = 2(a+b)$$

and

$$A = ab$$

So that

$$F_M = \frac{2}{3}\bar{c}\frac{a^2 b^2}{(a+b)\ell} \tag{2.17}$$

With the help of (2.11) and (2.16) we can contrast the conductance at high pressure, F_V, where viscous flow occurs, and low pressure, F_M, where molecular flow occurs

$$\frac{F_V}{F_M} = \frac{3}{16}\frac{p_a a}{\eta c} \tag{2.18}$$

and note that the conductance of a cylindrical tube is always greater under viscous flow conditions than it is under molecular flow. This arises because we have no control over molecular flow; it is determined by pipe geometry, whereas viscous flow can be driven to some extent by increasing the pressure differential and hence p_a. Examination of (2.16) shows that for high conductance in the molecular flow region we should choose short pipes of large diameter, although we must exercise some caution, as will become clear in our consideration of short pipes.

2.5 EFFUSION AND MOLECULAR FLOW THROUGH SHORT TUBES; THE CLAUSING COEFFICIENT

Consider an isothermal vessel, Fig. 2.3, which is divided by a thin partition pierced by a hole of area A, through which gas can escape from chamber 1 at pressure p_1 to chamber 2 at pressure p_2 and vice versa, $p_2 > p_1$. Both p_1 and p_2 are assumed to be low pressures.

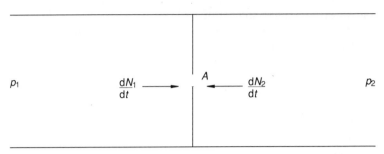

Fig. 2.3 Molecular flow through an orifice of area A, shown in section.

The rate at which gas molecules collide with the hole or orifice from chamber 1 is dN_1/dt, whilst in the reverse direction it is dN_2/dt. The net flow rate dN/dt is just the difference between these two flow rates. Using (1.48) we can say that

$$\frac{dN_1}{dt} = \frac{n_1}{4}\bar{c}A \tag{2.19}$$

$$\frac{dN_2}{dt} = \frac{n_2\bar{c}A}{4} \tag{2.20}$$

Substituting for n_1 and n_2 using (1.9) we obtain

$$\frac{dN}{dt} = \frac{\bar{c}A}{4k_B T}(p_2 - p_1) \tag{2.21}$$

so that we can write

$$Q_F = \frac{\bar{c}}{4}A(p_2 - p_1) \tag{2.22}$$

and the orifice conductance, F_o, is then given by

$$F_o = \frac{\bar{c}A}{4} \tag{2.23}$$

It should be noted that this result only applies if the following conditions are met:

1. The orifice diameter must be $\leq \frac{1}{10}\lambda$ where λ is calculated for the pressure p_2, the higher pressure.

2. The orifice must not be so large that mass transport of gas occurs.
3. The wall containing the orifice must be vanishingly thin immediately around the orifice; we are really considering a hole of zero length.

If we calculate the ratio of the conductance, F_o, of an orifice of radius a to that, F_p, of a pipe or tube of radius a we find that

$$\frac{F_o}{F_p} = \frac{3}{8}\frac{\ell}{a} \qquad (2.24)$$

This result directs our attention to an interesting aspect of molecular flow through tubes, that is that the inlet aperture to a tube has an impedance to gas flow which is significant compared with that of the tube itself, particularly if the tube is not long compared with the tube radius, a.

A simple, but approximate, way of handling this problem, originally due to Dushmann (1922), is to consider that the overall conductance may be obtained by using (2.4), thus we combine the orifice conductance F_o and the pipe conductance F_p to get

$$\frac{1}{F_M} = \frac{1}{F_o} + \frac{1}{F_p} \qquad (2.25)$$

We saw that, on the face of it, (2.16) suggests that we should use short pipes of large diameter to achieve maximum conductance, but (2.24) shows that once we lose the concept of a long pipe we can easily reach the situation where the impedance, Z_o, of the open end of the pipe becomes greater than that of the pipe itself, Z_p. So, for example, if we choose ℓ/a to be unity, then $F_o = \frac{3}{8}F_p$, and $Z_o = \frac{8}{3}Z_p$; the impedances become essentially equal when the pipe length is about three times its radius. The importance of the conductance of the aperture to a short pipe can more easily be understood by remembering that, under molecular flow conditions, there is no way in which we can 'steer' molecules into the pipe. Entry is purely random, but once in, flow is guided to some extent by collisions with the walls.

Equation 2.25 can be reset to take account of the end effect by writing

$$F_M = K'A\frac{\bar{c}}{4} \qquad (2.26)$$

where

$$K' = \frac{1}{(1 + 3\ell/8a)} \tag{2.27}$$

In fact, K' can be thought of as representing the ratio between the rate at which gas leaves the outlet of the tube and that at which gas strikes the inlet, i.e. the outlet probability. This approximate, but simple, approach of Dushmann to the problem of allowing for the effect of end conductance was examined more carefully by Clausing (1932), who produced the relationship

$$F_M = K_c A \frac{\bar{c}}{4} \tag{2.28}$$

The Clausing coefficient, K_c, is a dimensionless function of ℓ/a which ranges from unity, for $\ell/a = 0$, to $8a/3\ell$, for ℓ/a very large. It differs slightly in magnitude from the value of K' obtained by the simple approach of Dushmann and, as one would expect, approaches K' for large values of ℓ/a. Some values of K_c are plotted against ℓ/a in Fig. 2.4.

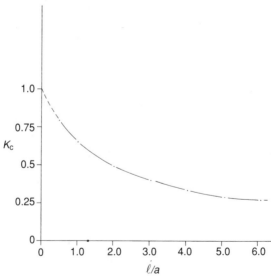

Fig. 2.4 A plot of Clausing's factor K_c against ℓ/a.

The seriousness of neglecting the 'end correction' when evaluating the conductance of tubes in the molecular flow region can be seen by inspection of Fig. 2.5 which shows the ratio, $\mathfrak{R}$, of the conductance of a 'long tube', calculated from (2.16), to that for the same tube using Clausing's correction factor K_c. The value of $\mathfrak{R}$ is essentially equal to unity when $\ell/a = 100$ (the 'long tube' condition). Otherwise the orifice imposes a significant conductance decrease. If one is dealing with short tubes of non-cylindrical form, the approximate approach of Dushman, exemplified by (2.25), may still be employed. In some of these cases, values of Clausing's K_c factor are available.

The expression for flow conductance in the molecular flow range is attractively simple and often easy to apply. It should be noted, however, that it depends on $\bar{c}$ and hence $T^{1/2}$ and $M^{1/2}$. As a consequence, the conductance of a pipe will alter dramatically as the nature of the gas being pumped is changed, thus, the conductance for hydrogen will be

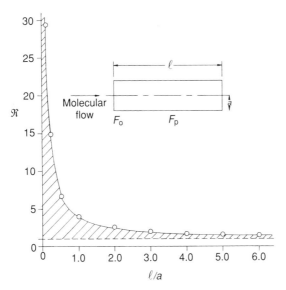

Fig. 2.5 Plot of the ratio $\mathfrak{R}$ for molecular flow in a circular section tube of length ℓ and radius a, where $\mathfrak{R}$ equals tube conductance/tube conductance calculated using Clausing's factor K_c. It can be seen that as the tube is shortened, the conductance calculated using Clausing's factor, which takes account of the orifice impedance F_o, is much less than that obtained from the straightforward calculation. There is essentially no difference when $\ell/a \geq 100$.

√14 times greater than that for nitrogen. This effect can sometimes be made use of in leak detection. The reader is referred to Chapter 8. Further, if the temperature is reduced from room temperature to 77 K (liquid nitrogen temperature), as is usual in a cold trap, then the conductance, for any gas, is halved.

2.6 FLOW IN THE TRANSITION RANGE; KNUDSEN FLOW

It has been suggested previously that we might expect flow in the transition region to possess characteristics lying between those for viscous and molecular flow. Equation (2.11) predicts that, for viscous flow, flow conductance goes to zero as p_a, the average pressure, goes to zero Under conditions lying outside those for strictly viscous flow, as defined in conditions (1 – 4) of section 2.3, we cannot expect to see this behaviour. Instead we should expect to see the concept of 'slip' manifest itself again as condition (3) of section 2.3 is violated and the flow velocity at the tube walls ceases to be zero.

In fact, if the pressure is reduced to such an extent that the Knudsen number is greater than 0.01, measurements indicate that extrapolation of the pressure, p_a, to zero actually yields a finite conductance, as shown in Fig. 2.6. This excess conductance can be attributed to the slip effect and considered as an additional term in (2.11). So, as pressures are reduced even further below the viscous limit the contribution from 'slip' increases, and the flow characteristics begin a progressive change from those appropriate to viscous flow to those characteristic of molecular flow (the changes taking place over about two orders of magnitude change in pressure). Clearly, there will come a point where the viscous characteristics of the flow are entirely absent, yet the flow is not strictly describable by the molecular flow equations. Despite this, the complete transition region is frequently identified with slip.

If the viscous conductance is corrected for slip we can use a result derived by Kennard and write

$$F_v = \frac{\pi a^4}{8\eta\ell} p_a \frac{(1+4\zeta)}{a} \qquad (2.29)$$

where ζ is the slip coefficient defined by (1.22); F_V is plotted in Fig. 2.6. Careful choice of a value for θ, the diffuse reflection coefficient in the expression for ζ, ensures that (2.29) predicts the slip conductance correctly. A value of θ between 0.8 and 1.0 is appropriate.

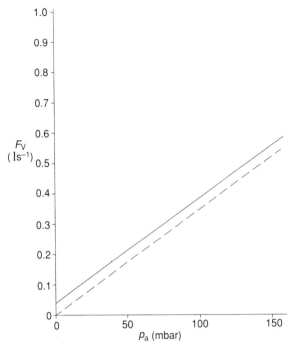

Fig. 2.6 The viscous conductance F_V of a long tube ($\ell/a \geq 100$) as a function of mean pressure p_a, in the viscous flow region, $\lambda/a < 0.01$. The broken line is the result calculated using Poiseuille's equation. The full line is the result of correcting for slip (data are for CO_2 and due to Knudsen).

Equations (2.29) and (1.22) may be combined and re-expressed in terms of the viscous (or slip-free) conductance F_V. Then the molecular flow conductance represented by (2.11) may be incorporated with a suitable factor. The result is applicable to long tubes and is given by

$$F = F_V + \frac{3\pi}{16} \frac{(2-\theta)}{\theta} F_M \qquad (2.30)$$

This equation is attractive in as much as it shows characteristics of both the viscous and molecular flow regions, as required. Unfortunately, in practice it does not predict the pressure dependence observed experimentally. A better approach, due to Knudsen, is an entirely empirical relationship of the same form as (2.30), but which can be written as

$$F = F_V + XF_M \qquad (2.31)$$

where X is a function of the ratio a/λ_a. In this instance λ_a is the mean free path corresponding to the average pressure p_a. Values for X may be calculated from the equation

$$X = \frac{1 + 2.507(a/\lambda_a)}{1 + 3.095(a/\lambda_a)} \qquad (2.32)$$

The function X ranges from a limiting value of 0.81 at the high pressure end of the intermediate flow region, up to unity at the low pressure end. At the low pressure end the term F_V becomes negligible,

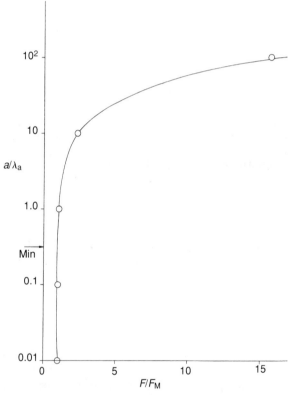

Fig. 2.7 A plot of the variation with pressure, actually a/λ_a, of the ratio of the total conductance, F, to that due to molecular flow alone, F_M.

as a consequence of its dependence on p_a, and $F = F_M$. The variation with pressure is shown in Fig. 2.7 where the ratio F/F_M is plotted against a/λ_a.

If (2.31) is required specifically for the transition region, Fig. 2.8, it is found to yield a minimum in the conductance/pressure relationship, although one might have expected a smooth increase in conductance between the molecular and viscous flow regions. It has been suggested by Pollard and Present (1948) that the minimum occurs due to competition between two different processes. As the pressure rises from the molecular flow range, initially the gas flow decreases as the long diffusion paths of the molecules are obstructed by the added molecules. With further increase in pressure and gas density, it becomes possible to establish some sort of drift motion which actually increases the conductance and is the precursor to viscous flow proper.

Now the obstruction of straightforward diffusion is going to depend on the ratio ℓ/λ. To establish some sort of drift motion we need intermolecular collisions, so it is useful to look at the ratio between the gas collision frequency and the wall collision frequency. This is equal to

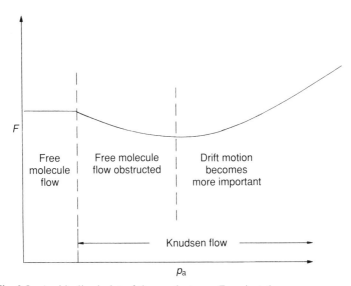

Fig. 2.8 An idealized plot of the conductance F against the average pressure p_a, in the pressure range up to 1 mbar. The constant conductance exhibited in the molecular flow region goes through a shallow minimum in the transitional or Knudsen flow region.

$2a/\lambda$ for unit length of tube. To summarize, the **decrease** in conductance due to the obstruction of diffusion depends on ℓ/λ and is therefore important primarily at low pressures where λ is large. There is a competing drift motion which develops with increasing pressure and depends on a/λ so that it manifests itself at higher pressures where λ is comparable with a. The development of drift motion leads to an **increase** in conductance which, in turn, develops into true viscous flow as the pressure is raised beyond that required to meet the condition $\lambda/a = 0.01$.

2.7 FREE-MOLECULE CONDUCTANCE OF TUBES IN SERIES

The effective conductance F of two tubes in series has been stated in (2.4) to be equal to $1/F_1 + 1/F_2$. In fact, this equation is not generally true. The error arises from the fact that it is seldom possible to specify the pressure at the tube junction when one is operating in the molecular flow regime. This is because the number of molecules crossing the junction in the two directions may be quite different when the mean free path is much larger than the tube dimensions. Under these conditions the effective pressure at the junction becomes a function of direction and has no unique value.

The result expressed by (2.4) will only be realised if the tubes are connected by large diameter sections in which the concept of junction pressure has some meaning. Such an arrangement is shown in Fig. 2.9(a). Otherwise, an approach originally due to Oatley (1957) may be adopted, which uses the Clausing coefficient introduced in the previous section. Since the Clausing coefficient may be regarded as the ratio between the rate at which gas emerges from the outlet of a pipe and the rate at which it enters the pipe, we can specify the conductance of tubes in series by means of an expression involving the respective Clausing coefficients and (2.28).

If we consider two tubes A and B of identical radius, joined coaxially, Fig. 2.9(b), and having Clausing coefficients K_A and K_B, then we can write, following Oatley, that the effective Clausing coefficient K_c is given by

$$\frac{1}{K_c} = \frac{1}{K_A} + \frac{1}{K_B} - 1 \qquad (2.33)$$

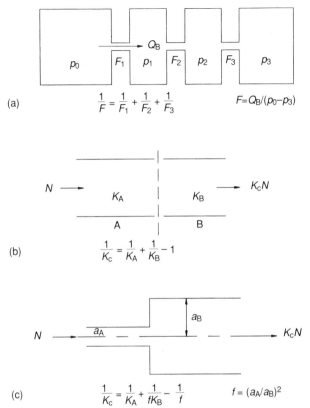

(a) $$\frac{1}{F} = \frac{1}{F_1} + \frac{1}{F_2} + \frac{1}{F_3}$$ $F = Q_B/(p_0 - p_3)$

(b) $$\frac{1}{K_c} = \frac{1}{K_A} + \frac{1}{K_B} - 1$$

(c) $$\frac{1}{K_c} = \frac{1}{K_A} + \frac{1}{fK_B} - \frac{1}{f}$$ $f = (a_A/a_B)^2$

Fig. 2.9 Free-molecule conductance of tubes in series. (a) The situation here is described by the straightforward application of the series formula for pipe conductances. Note the large diameter interconnecting regions which allow junction pressure to be defined. (b) The use of the Clausing coefficient for pipes of identical diameter joined coaxially. (c) The use of the Clausing coefficient for non-identical tubes joined coaxially.

For the more general case where more than two tubes of identical radius are connected in series we can write

$$\frac{1 - K_c}{K_c} = \frac{1 - K_1}{K_1} + \frac{1 - K_2}{K_2} + \ \cdots \ \frac{1 - K_n}{K_n} \tag{2.34}$$

Here we have n tubes in series with Clausing coefficients of $K_1, K_2, \ldots K_n$ respectively.

However, if two tubes of differing radii, Fig. 2.9(c), are involved, then (2.33) must be modified to take account of the fact that the conductance of the combination must be the same for gas flowing from left to right, or right to left. If this were not so we could conceive of using this structure to join together two containers initially at equal pressure and we would then find that the pressure would rise in one container and fall in the other; clearly nonsensical. The condition we require is embodied in (2.28), namely, that

$$K_A a_A^2 = K_B a_B^2$$

so we may rewrite (2.33) as if we were dealing with tubing of overall diameter a_B throughout, in which case K_A is replaced by K_A/f, where $f = (a_B/a_A)^2$ is the ratio of the cross sectional areas. The overall Clausing factor K_c' in this case is given by

$$\frac{1}{K_c'} = \frac{f}{K_A} + \frac{1}{K_B} - 1 \qquad (2.35)$$

Alternatively, and equivalently, since K_c must equal fK_c' we can write (2.35) in terms of tubing of radius a_A, thus

$$\frac{1}{K_c} = \frac{1}{K_A} + \frac{1}{fK_B} - \frac{1}{f} \qquad (2.36)$$

2.8 FLOW IN VACUUM SYSTEMS; THE SPEED OF A PUMP

The preceeding sections have looked in some detail at the description of gas flow as a function of pressure. In all vacuum systems, gas flow is a consequence of the action of a pump of some sort and it is appropriate at this point to define some of the quantities of importance in describing a vacuum pump and how these quantities are altered by connection to a vacuum system.

We are basically interested in the volume of gas that flows into the pump in unit time at a constant pressure. This is known as the intrinsic speed of the pump, S_p. Usually S_p is measured near the pump inlet where the pressure is p_i. In SI units, S_p will be expressed in $m^3 s^{-1}$, although units of ls^{-1} or $m^3 h^{-1}$ are also often found. The pumping speed of many pumps is approximately constant over a wide range of pressures.

The flow rate, Q_p, into the pump is known as the throughput of the pump; it decreases as the pressure drops and is usually specified at room temperature. We can write

$$Q_p = S_p p_i \qquad (2.37)$$

where Q_p will be expressed in Pa m^{3}s^{-1}, or mbar ls^{-1}.

It is apparent from the above definition that pumping speed has the same dimensions as conductance and we shall see that it may be used rather like conductance. Unlike conductance, however, pumping speed is not a property of a passive component such as a pipe, and of course, while conductance is throughput divided by the pressure drop across a tube, pumping speed is the throughput divided by the pressure in a specific plane.

In a vacuum system it is necessary to know the pumping speed at the inlet of a pump or at the end of a pumping line. Suppose that the pumping line has a conductance F, then the observed speed of the pump, S'_p, will be controlled by the conductance of this tube and we can write

$$\frac{1}{S'_p} = \frac{1}{S_p} + \frac{1}{F} \qquad (2.38)$$

It can be seen from this equation that for the observed pumping speed to be similar to S_p, F must be much greater than S_p. A ratio of 10:1 say, yields a value for S'_p of $(10/11)S_p$. The speed of pumps, and pump performance in general, is considered further in Chapter 4.

PROBLEMS – CHAPTER 2

2.1 A vacuum bell-jar is pumped by a pump having an observed pumping speed of 1000 ls^{-1}. The bell-jar contains a nitrogen ion source fed with N$_2$ gas from an external source. Ions are emitted into the bell-jar, from the ion source, through an aperture having a conductance of 2.004 ls^{-1}. If the measured gas pressure in the bell-jar is 10^{-6} mbar, what is the gas pressure inside the ion source?

2.2 Calculate, *approximately*, the conductance for molecular flow of N$_2$, for a tube having a length of 10 cm and a diameter of 20 cm.

2.3 Nitrogen gas is flowing, under molecular flow conditions, into a tube of radius 5 cm and length 10 cm (Clausing factor 0.80) which opens out abruptly into a tube of radius 10 cm and length 10 cm (Clausing factor 0.67). What is the overall conductance of this arrangement?

2.4 A cold trap has a molecular flow conductance, measured for nitrogen gas, of 500 ls^{-1}; it is connected between a vacuum chamber and a diffusion pump having a nitrogen speed of 100 ls^{-1}. Calculate the pumping speed at the entrance to the cold trap when the cold trap is cooled to liquid N_2 temperature (77 K) and the principal gas being pumped is H_2.

3

Physisorption, chemisorption and other surface effects

3.1 INTRODUCTION

Before examining in detail the operation of vacuum pumps for the production of vacua, it is important to consider a class of phenomena which can, on the one hand, impede the production of vacua or, on the other, may be utilized as the basis of special pumping systems. In any leakproof vacuum system there are basically two categories of gas to be considered: that which occupies the main volume of the system and which is readily removed simply by pumping, and that which is present on, or in, the surfaces of the system. It is this ability of surfaces to attract and store gas molecules which will concern us in this chapter; we shall find that the mechanisms which are involved in retaining gases on the inner surfaces of a vacuum system may also be used as the basis of three types of vacuum pump.

Where gas is taken up at the surface of a solid we distinguish two kinds of behaviour. In the first, the gas may enter into the solid in much the same fashion that a gas dissolves in a liquid: this process is known as **ab**sorption, or sometimes occlusion. Alternatively, the gas may simply interact with the surface of the solid: this process is known as **ad**sorption. In reality both types of behaviour are found together and no sharp boundary can be drawn between them. For the production of high, or particularly, ultrahigh vacuum (UHV) it is necessary to pay special attention to the material present at, or in, the vacuum system surface, and in this connection we note that the material which takes up the gas is known as the sorbent, adsorbent or absorbent, while the gas or vapour removed from the gas phase is known as the sorbate, adsorbate or absorbate. In respect of the production of high vacuum, we will be interested in the process of removing gas from a sorbent, a process known as desorption, or alternatively we may have occasion to use the sorptive properties to enhance a vacuum.

When a gas interacts with the surface of a solid it is obvious that its sorptive capacity depends largely upon the surface area per unit mass. Thus the sorptive capacity is much greater for porous substances than for smooth metals. The intermolecular forces which cause adsorption differ for different gas–solid systems. In some cases the nature and strength of the bond is akin to that found in chemical bonds and the process is then known as chemisorption. Chemisorption is associated with a high degree of adsorbent–adsorbate specificity and with large heats of adsorption; it tends to be either irreversible or reversible only with great difficulty.

Gases which have no chemical interaction with a solid may be adsorbed as a result of purely physical forces. These relatively weak interactions involve polarization forces or van der Waals forces and occur very generally. Because they are weak they tend to be most noticeable at low temperatures. The process involved is known as physisorption. Physisorption and chemisorption can occur together at low temperatures, with a physisorbed layer being formed upon a chemisorbed layer.

3.2 PHYSISORPTION

In physisorption, the adsorbate is bound to the surface by the van der Waals forces that arise from the interaction of fluctuating dipoles. The enthalpy (heat) of physisorption, or the binding energy of the adsorbate is, therefore, of similar magnitude to that of the corresponding heat of condensation of the gaseous adsorbate, $20\,kJmol^{-1}$, or less. One would expect that, since metals contain a large number of mobile conduction electrons, they might be expected to behave very differently from dielectrics. However, the resonance frequencies of the adsorbate atoms are so high that the conduction electrons are unable to adjust to the dipole fluctuations of the adsorbed atoms. In consequence, metals are incompletely polarizable and their adsorptive properties differ little from those of dielectrics.

Despite this it is found that the physisorption of inert gases on metals reduces the effective work function of the metal, which must imply a change in the surface distribution of conduction electrons. This change in the surface distribution of conduction electrons indicates that we should include an additional energy term along with the dispersion contribution going to make up the binding energy of the adsorbate. The usual approach to this problem is to consider that the adsorbed atoms are polarized by the large electric field present at the metal surface. If the volume polarization of the absorbed atom is α_p, then the energy required to produce these field induced dipoles is approximately $\frac{1}{2}\alpha_p^2 F_e$ where F_e is the field strength. As the atomic weight of the adsorbate increases, the contribution from this effect also increases, so that if we consider, for example, argon, krypton and xenon on, say, a tungsten surface, the heats of adsorption at low coverages are increased over the condensation heats by 1.3, 9.0 and 21.3 kJmol^{-1} respectively. The energetics of physisorption are depicted in Fig. 3.1 by curve ABFC which shows that the molecule will be bound to a metal surface with a heat of physisorption q_p at an equilibrium distance r_p.

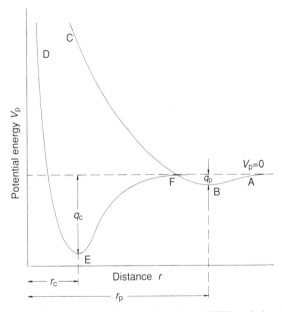

Fig. 3.1 Potential energy curves for physisorption (ABFC) and chemisorption (ABFED). The heat of adsorption for physisorption is q_p while that for chemisorption is q_c. The distances r_p and r_c represent the equilibrium bonding distances for physisorbed and chemisorbed species respectively. At the crossover point F, $V_p \sim 0$, showing that chemisorption in this instance is non-activated.

We can summarize the preceding account by noting that since physisorption heats are small, we shall only find substantial adsorption near, or below, the boiling point of the adsorbate, with the formation of multilayers requiring higher pressures. Also, the extent of adsorption on the surfaces of different solids under the same conditions of temperature and pressure will not vary much. Binding on metals will generally be stronger than that for non-metals, at least in the case of large, readily polarized molecules.

The usual way in which the sorption of a gas or vapour is determined for a particular solid is to begin with a surface which is free of all adsorbates. Because of the weak forces involved in physisorption, this situation is achieved simply by heating the adsorbent at a suitably high temperature in a good vacuum. Then, while the adsorbent is maintained at a constant known temperature, the sorbate gas is introduced into the system and a decrease in pressure is observed when equilibrium is attained.

For a given solid there will be a definite relationship between the amount of gas adsorbed per unit mass, or unit area of adsorbent, and the pressure. The amount of gas adsorbed will be some function of temperature and pressure. A plot of volume adsorbed against pressure at constant temperature is known as an adsorption isotherm. From adsorption isotherms at a series of temperatures, it is possible to plot adsorption isosteres which give the variation of the equilibrium pressure with temperature for a constant amount of gas adsorbed. The isosteres are thus analogous to the vapour pressure curves of liquids. From the isotherms it is also possible to plot isobars, which give the amounts adsorbed at a series of temperatures for a given constant pressure. Many of the earliest measurements were conducted on activated charcoal and examples are given in Fig. 3.2.

A number of algebraic expressions have been deduced to describe the variation in amount adsorbed with pressure at constant temperature. One of the earliest of these is the parabolic adsorption isotherm proposed by Freundlich in 1909. This is an empirical relationship which has the form

$$V = kp^{1/n} \qquad (3.1)$$

k and n are constants which depend upon the nature of both adsorbent and adsorbate together with the temperature. Equation 3.1 is generally expressed in logarithmic form as a plot of $\ln V$ against $\ln p$, which should yield a straight line of slope $1/n$. In all cases $n > 1$. Table 3.1 shows values for k and n obtained for the inert gases argon, krypton and

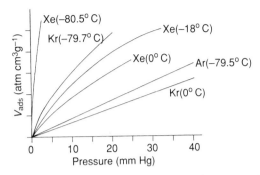

Fig. 3.2 Typical adsorption isotherms for argon, krypton and xenon on charcoal (Peters and Weil, 1930).

xenon adsorbed on charcoal. It will be noted that, for any one gas, the value of k decreases with increase in temperature, while the value of $1/n$ increases and ultimately reaches the value unity. Generally, k rises with rise in boiling point while $1/n$ decreases. For $n = 1$, the amount adsorbed varies directly with the pressure. This is the relationship embodied in Henry's law which describes the solubility of gases in liquids. It is also seen in physisorption at very low pressures.

The parabolic equation outlined above gives very little insight into the physisorption mechanism. Accordingly, in 1914 Langmuir developed a new view of the physisorption process which yields ultimately a hyperbolic adsorption isotherm. Langmuir's theory is very important in the sense that it was the first to introduce concepts like the 'monolayer', where the surface of the adsorbent is covered by a layer of adsorbed atoms in one to one correspondence, and the idea that the adsorption of one atom has no influence on the binding energy

Table 3.1 Values for k and n in the Freundlich adsorption isotherm

Gas	193 K			255 K			273 K		
	k	n	$1/n$	k	n	$1/n$	k	n	$1/n$
Argon	0.5	1.05	0.95	0.076	1	1.0	0.058	1.0	1.0
Krypton	2.93	1.41	0.71	0.497	1.13	0.88	0.34	1.0	1.0
Xenon	15.99	1.75	0.57	2.46	1.45	0.69	1.58	1.29	0.77

associated with adjacent adsorbed atoms (adatoms)/molecules, i.e. no interaction between adsorbed species. The essential assumptions of the Langmuir model may be listed as

1. The adsorbent is planar and the total number of adsorption sites is constant under all experimental conditions.
2. The adsorption is restricted to a monolayer.
3. On impact of the gaseous adsorbate molecule with an unoccupied or free site, it is adsorbed with zero activation energy.
4. An occupied site comprises one adsorbate molecule at a single site and collision of a gaseous adsorbate molecule with an occupied site is elastic (this will not always be true for physisorption, as multiple layers are possible).
5. Desorption of an adsorbed molecule occurs as soon as it has acquired sufficient thermal energy from the lattice vibrations of the adsorbent to equal the heat of adsorption.
6. At equilibrium the rate of adsorption on unoccupied sites equals that of desorption from the occupied sites.

Following the above assumptions we can write the rate of adsorption per second per unit area for an adsorbate–adsorbent system at equilibrium at temperature T and at a pressure p of gaseous absorbate, assuming that there are N_S sites per unit area and that N of these are occupied. The rate of adsorption per unit area of surface s^{-1} is then

$$\frac{dN}{dt} = k_a p (N_S - N) \tag{3.2}$$

where $N_S - N$ is the number of unoccupied sites and k_a is the rate constant per site for unit pressure; $k_a N_S$ equals the corresponding collision rate v given by the Hertz–Knudsen equation for unit pressure. Usually the capture rate on collision is less than unity so that $k_a = sv$ where s is the sticking probability per unoccupied site (see section 3.5), and the rate of desorption per unit area of surface per second is $k_d N$ where k_d is the rate constant per adsorbed molecule for desorption: it has the dimensions s^{-1}.

At equilibrium

$$k_a p (N_S - N) = k_d N \tag{3.3}$$

or

$$k_a p (V_m - V) = k_d V \tag{3.4}$$

where V is the volume of gas adsorbed and V_m is the volume to give a complete monolayer, both in, say, cm^3 at STP. Hence,

$$V = k_a p V_m / (k_a p + k_d) \tag{3.5}$$

Since the fractional coverage is $\theta = N/N_S = V/V_m$,

$$\theta = \frac{k_a p}{\left(k_a p + k_d\right)} \tag{3.6}$$

or

$$p = \frac{k_d}{k_a} \frac{\theta}{(1-\theta)} = K_a \frac{\theta}{(1-\theta)} \tag{3.7}$$

where $K_a = k_d/k_a$ is the equilibrium constant for adsorption. Equation (3.6) is the equation of a rectangular hyperbola, hence isotherms having this form are termed hyperbolic isotherms.

The validity of the hyperbolic representation may be tested by plotting p/V against p, which should yield a straight line with slope equal to $1/V_m$ since we can rewrite (3.7) as

$$\frac{p}{V} = \frac{p}{V_m} + \frac{K_a}{V_m} \tag{3.8}$$

The intercept on the ordinate axis for $p = 0$ is K_a/V_m. Alternatively, we can recast (3.8) in terms of $1/V$ alone so that a plot of V/p against V should yield a straight line with slope $-1/K_a$ and intercept for $V = 0$ of V_m/K_a.

At very low pressures where $p/K_a \ll 1$ and $V \ll V_m$

$$V = p V_m / K_a \tag{3.9}$$

which is merely a statement of Henry's law.

The derivation above assumes that adsorption is not accompanied by dissociation. However, if the molecules are adsorbed as single atoms the expression for the coverage, (3.6) must be altered to

$$\theta = \frac{\left(p/K_a\right)^{1/2}}{1 + \left(p/K_a\right)^{1/2}} \tag{3.10}$$

or

$$V = \frac{V_m \left(p/K_a\right)^{1/2}}{1 + \left(p/K_a\right)^{1/2}} \tag{3.11}$$

so that, even at relatively low pressures, the total amount of gas adsorbed varies in proportion to the square root of the pressure.

The hyperbolic equation produced by Langmuir has been found to give good agreement with experimental data in many adsorption studies. However, values obtained for the constants V_m and K_a are often difficult to reconcile with the physical situation. In many instances, where the value of the surface area is reasonably well-known, the value found from V_m, and the gas molecule diameter determined from viscosity measurements, is smaller than expected. Moreover, different gases yield different values for a given surface. This is particularly true where the weak forces of physisorption are involved; chemisorption is often better represented.

One of the possibilities existing in the physisorption process is the production of multilayer adsorption. Figure 3.3 shows the typical isotherm types found in physical adsorption: Fig. 3.3(a) shows the hyperbolic isotherm described by Langmuir; the other four represent the effects of multilayer adsorption. Perhaps the most useful theoretical representation of multilayer adsorption is that due to Brunauer, Emett

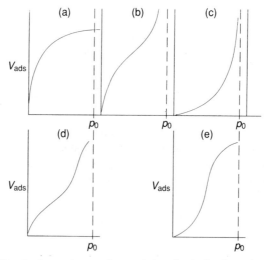

Fig. 3.3 The isotherm types observed in physical adsorption. Type (a) represents the Langmuir isotherm for monolayer adsorption. Types (b)–(e) represent multilayer adsorption.

and Teller (BET). It is based on the Langmuir model with the additional ideas that the rate of evaporation of the second layer is equal to the rate of condensation on the first layer, and so on for successive layers. Also, it is assumed that the heat of adsorption in the second and succeeding layers is equal to the heat of liquefaction of the bulk adsorbate, i.e. the van der Waals forces of the adsorbent surface affect the first layer of adsorbate only.

The equation which results for a free surface is

$$\frac{p}{V(p_0 - p)} = \frac{1}{V_S c} + \frac{(c-1)}{V_S c} \frac{p}{p_0} \tag{3.12}$$

where V_S and c are constants at any given temperature and p_0 is the saturation pressure of the gas at the given temperature (see Figs 3.3(b–e)). When $p = p_0$, the amount adsorbed is infinitely great. Since its introduction, the BET equation (3.12) has found widespread use as a means of determining surface areas. The adsorption isotherms for nitrogen and the inert gases argon and krypton, measured at liquid nitrogen temperature, 77 K, fit the BET equation over a wide range, so that if the isotherm is measured over a range of pressures up to $p/p_0 \simeq 0.3$, and plotted according to (3.12) as $p/V(p_0 - p)$ versus p/p_0, the resulting linear plot should have slope $(c-1)/V_S c$ and intercept $1/V_S c$. Here V_S is the volume adsorbed in the first monolayer; c is a constant, approximately equal to exp $[(E_1-E_L)/RT]$, where E_1 is the average heat of adsorption in the first layer and E_L is the bulk heat of liquefaction. In general, heats of adsorption for the same gas on different adsorbents are roughly the same and the heats increase with increase in the boiling point of the gas. If one can assign a value to the area of a single molecule, then V_S tells one how many molecules are required to achieve a monolayer and hence the surface area may be determined.

Since experimental isotherms of many different shapes exist, it is possible to find one, or several, which can be represented satisfactorily within a range of equilibrium pressures by almost any selected equation, although if the range is too wide hardly any popular equation will account for the isotherm. In view of the fact that the equations discussed involve two adjustable constants, it may happen that a large portion of an experimental curve can be satisfied by more than one equation and thus accord between theory and experiment does not necessarily demonstrate the validity of the model used to produce the equation.

3.3 CHEMISORPTION

Chemisorption, as was pointed out in the introduction to this chapter, is associated with a high degree of adsorbent–adsorbate specificity. It may be considered to be a chemical reaction between an adsorbate molecule and a surface atom, where the atom involved will be a metal atom. The greatest interaction is with metal atoms in immediate contact with the adsorbed molecules, but the weaker interactions with all other surface atoms cannot be ignored. The binding energies, or heats of adsorption, are in the range from 50–$400\,\mathrm{kJmol^{-1}}$, and are thus similar in magnitude to those found in the chemical bonds of free molecules. Figure 3.1 shows the potential energy of a chemisorbed atom as a function of distance from the adsorbent surface.

The short range nature of the forces involved in chemisorption generally means that adsorption is confined to the monolayer, but this does not preclude the formation of an additional weakly chemisorbed, or perhaps physisorbed, layer on top under the appropriate conditions of temperature and pressure, i.e. low temperature and higher pressures.

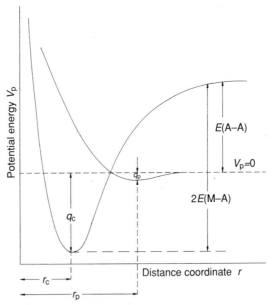

Fig. 3.4 Potential energy curves for dissociative chemisorption with zero activation energy. $E(\text{A–A})$ represents the dissociation energy of the free molecule; $2E(\text{M–A})$ is twice the bonding energy of the atom A to the metal surface.

Generally, chemisorption is well described by the Langmuir model. There is another aspect of chemisorption which distinguishes it from physisorption, and that is the occurrence of dissociative chemisorption. Here a gaseous adsorbate molecule is dissociated into its component atoms and/or radicals. The adsorbed species then comprise adsorbed atoms (adatoms) or/and radicals, Fig. 3.4; where the adsorbed species does not differ from the gaseous molecule we have associative chemisorption. It is possible for both dissociative and associative chemisorption to occur on the same metal although associative chemisorption will be favoured by low temperatures.

The driving force for dissociative chemisorption is the presence of adjacent, unsaturated metal bonds which are able to react with the adsorbate molecule to lead to a situation where the energy gained from forming the two metal–adsorbate atom bonds exceeds the dissociation energy of the free molecule. There are instances, however, where the passage from the free molecule into the chemisorbed state requires a small amount of energy, the activation energy, as shown in Fig. 3.5. Generally, the chemisorption of most gases on metals is a non-activated

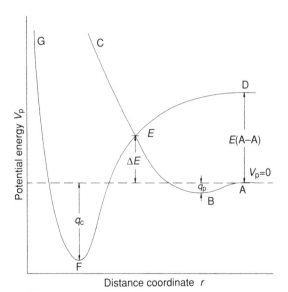

Fig. 3.5 Potential energy curves for dissociative chemisorption with an activation energy ΔE. Curve (ABEC) represents physisorption; (DEFG) represents chemisorption; $E(A–A)$ represents the dissociation energy of the free molecule.

process so that the process is controlled by thermodynamics rather than by kinetics. An exception is the chemisorption of nitrogen by iron, which requires a substantial activation energy.

It has been implied in the foregoing account that all metals will participate in chemisorption; this is not generally true. The transition metals are particularly active in chemisorption and both their electrical and magnetic properties are altered in the process. In consequence, chemisorption activity has been correlated with the presence of unpaired d-electrons in the metal and their participation in the formation of a strong chemical bond. The coinage metals, copper, silver and gold for example, have an sp character and the metal–adatom bond is weaker, so much so that it is insufficiently large to effect dissociative chemisorption. Despite this, hydrogen atoms do bind strongly to these metals at low temperatures with an energy similar to that found for chemisorption on transition metals. Usually it is necessary to provide some means to predissociate the hydrogen molecule, e.g. atomize them at a hot tungsten filament such as is found in an ionization gauge. The behaviour of copper and gold is different with respect to hydrogen atom

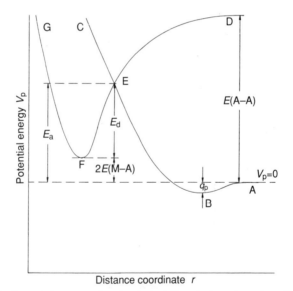

Fig. 3.6 Potential energy curves typical of dissociative adsorption of hydrogen on gold where $E(A–A)$ is the dissociation energy of the free hydrogen molecule; E_d is the energy of desorption as hydrogen molecules and E_a is the activation energy for chemisorption as hydrogen atoms. Curve ABEC is the curve for physisorption of hydrogen molecules.

adsorption since, on gold, the potential energy of the two adsorbed hydrogen atoms is higher than that of a free hydrogen molecule but lower than that of two gaseous hydrogen atoms, whereas on copper the reverse is true (Figs 3.5 and 3.6). Consequently, hydrogen atoms adsorbed on gold at low temperatures become mobile as the temperature is increased and when two atoms collide they are desorbed as a molecule.

3.4 THE CONDENSATION COEFFICIENT

When a molecule approaches a surface it experiences an attractive potential which accelerates it towards the surface, increasing the velocity component perpendicular to the surface. The larger the binding energy involved in the subsequent bond the greater the acceleration and, of course, the deeper the potential well. In the collision between the molecule and the surface, the incoming molecule must transfer to the lattice an amount of energy equal to, or greater than, its translational energy perpendicular to the surface if condensation is to occur. The efficiency of this energy transfer process increases as the molecule binding energy increases, since the collision time is now longer. During the collision, a compressional wave is propagated through the lattice or, put another way, energy is absorbed by phonon excitation in the metal lattice.

The efficiency of this process is described by the condensation coefficient, and for most adsorbates on metals it approaches unity. However, although the adsorbate molecule has been captured, it may well be in an excited state and it will still retain tangential velocity components, which means that thermal equilibrium with the surface has not been established and a final adsorption site is not yet occupied. The condensation coefficient basically describes the initial stages of adsorption, i.e. capture in the van der Waals potential well; the molecule is thus physisorbed at this point in time. This process precedes the final act of chemisorption for which the capture efficiency is described as the sticking probability which is in turn measured by the sticking coefficient.

3.5 THE STICKING PROBABILITY

It might be thought that the condensation coefficient and the sticking coefficient would be identical, but this is not so. The sticking probability is defined as the ratio of the rate of capture of a molecule into the chemisorbed state to the rate of collision of the gaseous molecules with the surface. Capture into the chemisorbed state requires the presence of an unoccupied adsorption site, so that the transition from the weakly bound physisorbed molecule to the strongly bound chemisorbed molecule/atom will occur rapidly if the initial collision is with an unoccupied site, otherwise the physisorbed molecule will desorb. The sticking probability will, therefore, almost always be less than the condensation coefficient. If the collision occurs with an already occupied site, only weak physisorption occurs with the result that the molecule returns to the gas phase with some loss of kinetic energy. Consequently the sticking coefficient decreases, approximately linearly, with increasing coverage.

It may happen that the colliding molecule impacts with a chemi-sorbed species in such fashion that a weakly bound precursor state is formed. This precursor state (the precursor to chemisorption) will generally be mobile and may then be transformed to the chemisorbed state before desorption occurs. Under these circumstances the sticking probability will be substantially independent of coverage over a wide range of coverages. This situation is favoured by low coverages and low temperatures.

The value of the sticking coefficient for many gases on a wide range of metals, particularly transition metals, has been measured. The values obtained are generally of the order of 0.1, but can reach as high as ~ 1; values are recorded in Table 3.2 for some common metals.

Table 3.2 Initial sticking coefficients for some common gases and metals[*]

Gas	Ti	W	Ni
H_2	0.06	0.5	0.3
CO	0.7	0.4	0.9
N_2	0.3	0.95	–
O_2	0.8	0.98	0.95
H_2O	0.5	0.1	0.2
CO_2	0.5	0.95	–
He	–	–	–
Ar	–	–	–
CH_4	0.05	0.1	–

[*] Values are for 300 K and must be regarded as nominal; they depend sensitively on the crystallographic character of the metal surface.

The actual value of the sticking coefficient is of considerable significance since, in essence, it determines the absorption of gases by metal surfaces and this property lies at the heart of vacuum production by 'getters', where a getter is a metal surface which pumps by chemisorption. We shall see that physisorption may also be used as a means of vacuum production and forms the basis of sorption pumping or cryopumping systems: the condensation coefficient, however, is essentially unity in physisorption for all gases and therefore may be ignored.

3.6 GETTERS AND GETTERING

Getters and gettering have been used, since the early twentieth century, to produce and maintain a low pressure in electron tubes (radio valves). They now find a more widespread application in the production of vacua through the use of the getter-ion pump. Although many metals, and particularly transition metals, will act as a getter, it is principally titanium which is used in this role nowadays. It getters, or chemisorbs, all common gases except methane. In some instances it is necessary to getter selectively as part of a purification routine. For example, rhodium or nickel will act as purifying getters for nitrogen since they absorb H_2, CO and O_2 but not N_2. Similarly gold absorbs CO and simple hydrocarbons but does not adsorb H_2, N_2 and O_2.

For a metal to function as a getter it must be presented to the vacuum system as an atomically clean surface. This is usually done by evaporating a metal film from a suitable wire source onto the surrounding vacuum envelope. Films formed in this way are

1. initially atomically clean,
2. of very high surface area, i.e. microporous.

Simple getter sources are depicted in Fig. 3.7. Figure 3.7(a) shows a section through a very versatile getter source formed by winding two equal diameter wires, one of tungsten, the other of the chosen getter material, around a tungsten core. The wires are wound with all turns touching. Passage of an electric current through this arrangement causes preferential evaporation of the chosen getter material. It is often possible, if due care is exercised, to evaporate from a simple wire helix, Fig. 3.7(b), made from the chosen material, although there are a number of metals with which this cannot be done, namely copper, silver, gold and aluminium. In the case of these latter materials, they melt before significant evaporation has occurred so that either the arrangement shown in Fig. 3.7(a) must be used or, more simply, a few turns of the selected metal in wire form is wound around the vee-section of the heater arrangement shown in Fig. 3.7(c). This comprises a tungsten wire heater which, when the melting point of the chosen material is reached, causes the material to congeal as a sphere suspended at the base of the 'vee' of the wire. The now molten sphere of metal provides a stable, point-evaporating source.

There are now commercially available sources of a wide range of metals, including the alkali metals. These latter sources are completely inert until activated. They provide linear sources of, for example, sodium or potassium in a most convenient way. The experimental arrangement is depicted in Fig. 3.7(d).

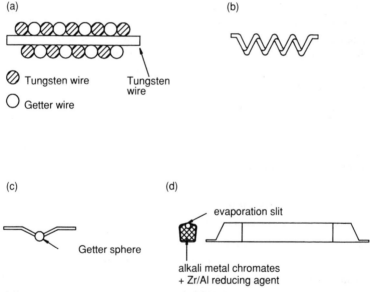

Fig. 3.7 Typical getter sources. (a) Tungsten wire overwound with equi-diameter tungsten and getter wires. (b) Helix of getter wire. (c) 'Vee' support for molten sphere getter sources. (d) Encapsulated source for alkali metals, produced by reducing the appropriate alkali metal chromate with Zr 84%, Al 16%, getter alloy. The reducing material getters any chemically active gases evolved during the reduction process. Very pure alkali metal is evolved from the longitudinal slit.

For significant pumping action the provision of an atomically clean surface must be maintained for long periods of time. This cannot always be done by the simple evaporation techniques outlined above, which have strictly limited capacity; recourse has to be had to an alternative technique for clean surface preparation using bulk material rather than a wire. This process is known as sputtering and the material used is again titanium.

3.7 SPUTTERING

Sputtering is a process in which surface metal atoms, titanium in this instance, are removed by bombardment with energetic ions and atoms. These energetic projectiles are formed in a discharge initiated above a titanium cathode and in the case of the ions are accelerated towards the titanium surface where they are able to couple their energy into the metal lattice. The effect of this energetic bombardment is to discharge surface titanium atoms by recoil forces from lower atomic layers, this process being particularly efficient if the angle of incidence is non-normal.

If we consider an energetic bombarding ion of mass M_1 and kinetic energy E_1 striking a surface made up from atoms of mass M_2, then application of the laws of conservation of energy and momentum show that the maximum energy E_{max} which can be acquired by the stationary atom is given by

$$E_{max} = \frac{4M_1M_2}{(M_1 + M_2)^2} E_1 \qquad (3.13)$$

Figure 3.8(a) sketches the situation described by this equation and shows that the momentum of the bombarding ion is coupled to a surface atom and directed into the surface. For the target atom to be ejected (sputtering), momentum reversal must occur. The necessary momentum reversal occurs if the incident ion penetrates into the target where it collides with, and is reflected from, a lower atomic layer in a sequence of collisions, as it gives up its energy to the target. Figure 3.8(b) shows why this momentum transfer and reversal process is more efficient if the angle of incidence is non-normal. If the incident ion has low energy (< 1 keV) then the mean energy $\bar{E}$ of the target atom is

$$\bar{E} = (E_{stat} + E_d)/2 \qquad (3.14)$$

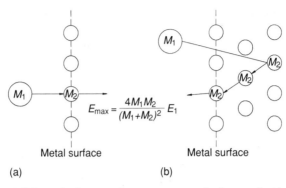

$$E_{max} = \frac{4M_1 M_2}{(M_1 + M_2)^2} E_1$$

Metal surface   Metal surface

(a)   (b)

Fig. 3.8 (a) Schematic diagram of momentum transfer from an incident ion to a metal surface. (b) Recoil action during ion impact resulting in metal atom sputtering.

where E_d is the energy required to displace a target atom and E_{stat} is the kinetic energy of the initially stationary atom after the collision, $E_{stat} \le E_{max}$. When the energy of the bombarding ion is too high, it buries itself very deeply in the target lattice and sputtering does not occur. This situation cannot arise when sputtering is initiated by a glow discharge since the potential drop across the cathode dark space will not usually exceed 500 V (Fig. 3.11).

For sputtering to occur efficiently, the number of atoms ejected per bombarding ion needs to be as high as possible; this quantity is known as the sputtering yield, S. Once the metal atom has its energy reduced by interatomic collisions to a value less than E_S, the surface binding energy, sputtering will not occur. The number of collisions N made in slowing down to this energy is

$$N = \frac{\ln(E / E_S)}{\ln 2} \qquad (3.15)$$

assuming one collision per interatomic distance. Now the total number of atoms displaced towards the surface per primary collision is $\bar{E}/2E_d$ because only half of the displaced atoms can migrate towards the surface. It has been shown that the number of atomic layers which contribute to sputtering is given by $1+\sqrt{N}$ so that we can write that

$$S = \frac{\bar{E}}{4E_d} \left\{ 1 + \left[\frac{\ln(\bar{E} / E_S)}{\ln 2} \right]^{1/2} \right\} \sigma_e n^{2/3} \qquad (3.16)$$

where n is the number of atoms of the target material per unit volume and σ_e is the ion collision cross section, a function of the ion energy. The foregoing analysis has stated that sputtering will occur more efficiently if the ion is incident on the metal (titanium) surface non-normally. However, if the ions are generated in a glow discharge then the cathode field orientation will necessarily mean that ions impinge on the cathode normally. To get round this problem, it is usual practice to slot the cathode surface to ensure that many ions will strike it obliquely, with high sputtering efficiency. This aspect of sputtering is described more fully in section 4.8. In any event, the sputtering coefficient will depend on the nature of the metal and the incident ion, its energy and the angle of incidence of the ion; it can take values which range from negligible up to about ten.

The titanium atoms released by this bombardment process are deposited as a gettering surface on adjacent electrodes, usually the anode in the case of slotted cathode designs. This deposition process not only provides fresh getting surface, but also permits the burial of inert gas ions previously accelerated to the cathode surface. Because a momentum transfer process is involved, light molecules such as hydrogen or atoms such as helium are not efficient in sputtering. They will be pumped initially by ion burial. Hydrogen will then diffuse into the bulk of the titanium cathode and form a hydride.

3.8 SORBENTS AND MOLECULAR SIEVES

There are three sorbent materials which have seen widespread use in vacuum production, namely, activated charcoal, activated alumina and molecular sieves. Activated charcoal, the oldest of the sorbents, is prepared by subjecting coconut shell charcoal to a heat treatment followed by baking *in vacuo* to drive out occluded gases. The resulting granular material has an effective surface area of about $2.5 \times 10^3 \, \text{m}^2\text{g}^{-1}$. This very high surface area per gram is typical of all the important sorbents, although the actual values vary from one type of material to another and depend on the exact activation process followed. Activated charcoal will absorb gases at room temperature, but sorption is of course greatly enhanced by lowering the temperature to 77 K by means of liquid nitrogen.

Activated alumina has exceptional trapping properties for oil vapours at room temperature and has a very long life in this role before it becomes saturated. It will typically have a surface area of $210 \, \text{m}^2\text{g}^{-1}$. In common with the other sorbents it readily absorbs moisture although, in this instance, it does not affect the oil vapour trapping properties and therefore makes this material ideally suited for use in foreline traps, as described in section 6.7. The full absorption properties are regenerated by heating to 250° C. It is prepared by the dehydration of böhmite, AlO(OH), or Al(OH)$_3$ at temperatures of about 450° C. The resulting material, γ Al$_2$O$_3$, exhibits a defect spinel structure where one ninth of the metal atom sites, in the 16 octahedral positions and 8 tetrahedral positions, are left empty at random. The material has very small particle size and very great adsorptive power resulting from its ability to take up gases not only in its surface, but also, in some instances, into empty metal sites.

The third category of sorbents are the zeolites or molecular sieves. Zeolites are hydrated alumino-silicates also containing sodium, potassium, calcium and barium, either singly or together according to the pore size required. Some of these minerals occur naturally but many are synthetic, particularly those which find use in vacuum work. When these minerals are outgassed at an elevated temperature, usually around 350° C, the water of hydration escapes but the crystal lattice does not collapse. Thus, the dehydrated crystal is a 'honeycomb' that is full of voids which had previously been filled with water molecules. The relative volume of these voids per unit volume is between 0.18 and 0.51. If a dehydrated crystal is placed in the presence of water vapour, the voids fill again with water molecules provided that the passages are sufficiently wide; the shapes of the passages can be determined from the crystal structure of the minerals. In these minerals, the whole **ad**sorbent is uniformly porous on the molecular scale and it is equally correct to state that zeolites **ad**sorb gases or that gases are **ab**sorbed by them. Indeed, in zeolites the occluded molecules occupy interstices left by the removal of H$_2$O from the original lattice structure and form a regularly disposed, often mobile, component. In some ways the process is better described as a solid solution.

The special importance of zeolites is that the space originally filled by water can be occupied by molecules of different gases. However, this is true only for certain gases since zeolites exhibit a second phenomenon known as persorption. Persorption may be defined as adsorption in pores only slightly wider than the diameter of the adsorbate molecules; some molecules will therefore be excluded. The nature of the material actually sorbed by a zeolite will be controlled by the diameter of the access channels joining the voids in the structure so that some gases, e.g. H$_2$O, CH$_4$, will perhaps be sorbed easily and rapidly while others (larger) will be sorbed slowly or not at all. It is a consequence of this selective sorptive action that these materials have gained the title 'molecular sieves'.

The principal zeolites in use are Linde sieves 4A, 5A and 13X. Linde sieve 4A and 5A conform to the general formula $Me_{12}[(AlO_2O)_{12} (SiO_2)_{12}]$. 27 H_2O, where Me is a metal ion, sodium in the case of Linde sieve 4A, or 4 Na^+ and 8 Ca^+ ions for Linde sieve 5A. The pore size will accept molecules of overall diameter up to 0.4 nm in the case where Na^+ ions only are present, or up to 0.5 nm if two thirds of the Na^+ ions are replaced with Ca^+ ions. These materials are totally synthetic and were developed by the Linde Air Products Co. in the USA. Figure 3.9 shows an outline of the structure of 4A and 5A. The important aspect to note is that the structure comprises a three-dimensional network of cavities interconnected by short channels. It is the overall diameter of these channels which determines the size of the molecules which may eventually reside in the cavities. If the temperature of the zeolite is reduced, the interconnecting channels contract, the amplitude of the atom vibrations is reduced and the kinetic energy of the gas molecules is reduced. Adsorptive ability for some gases, e.g. N_2, will therefore show a sharp decrease with temperature.

Linde sieves 4A and 5A have a cubic structure in which cavities ~ 1.1 nm in diameter are formed from cubo-octahedral blocks containing 24 (Si^{4+}, Al^{3+}) ions. These cavities have six interconnecting apertures formed from oxygen ions, where the effective diameter of the aperture can be adjusted by altering the compositions of the zeolite as described above. In Linde sieve 13X, the structure is made up from a

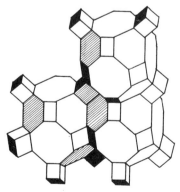

Fig. 3.9 Outline structure of Linde molecular sieves type 4A and 5A, showing cavities and interconnecting channels offering molecular storage.

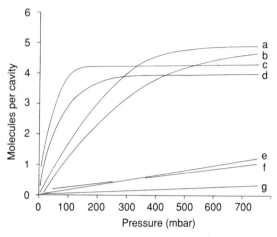

Fig. 3.10 Adsorption of saturated and unsaturated hydrocarbons by zeolite 5A at 150° C showing number of molecules in each cavity as a function of pressure. The unsaturated hydrocarbons are, top to bottom, (a) propylene, (b) ethylene, (c) acetylene and (d) iosbutylene; saturated hydrocarbons are (e) propane, (f) ethane and (g) methane.

diamond-like tetrahedral array of cubo-octahedra, each forming a cavity of ~ 1.3 nm diameter and joined by oxygen atoms forming four apertures with effective diameter of 0.9–1.0 nm. The sorption volume of this zeolite is 51%, the largest known for any zeolite; it has a surface area of $1030 \, \text{m}^2\text{g}^{-1}$. The sorptive ability of Linde sieve 5A is shown in Fig. 3.10 where the number of molecules 'stored' in each cavity is measured against pressure for some simple hydrocarbons. Molecules within these cavities tend to be held there by forces of electrostatic and van der Waals type so that we can expect that polar molecules of any sort, i.e. molecules exhibiting a permanent electric dipole such as water and carbon dioxide, or unsaturated hydrocarbons, will be held strongly provided that they are just small enough to enter the cavities. Molecules which are non-polar will be more weakly absorbed and this will be particularly true of very small molecules or atoms which can not only enter, but also leave, easily. In addition, weakly bound atoms or molecules may be displaced by more strongly bound polar molecules, thus water will usually act as a displacer on zeolites.

The bond strengths exhibited by absorbates in zeolite cavities are generally higher than are found on a free surface at the same temperature. This results from the very high electric fields at the surface of zeolite cavities, which leads to enhanced bonding energies through the mechanism of field-induced adsorption. Increases of bond strength by a factor of 50 are possible in the presence of a very high electric field.

For all three sorbents, examination of the appropriate isotherms shows that more gas can be absorbed at a given pressure if the temperature is reduced, simply because the probability of desorption decreases with decreasing temperature. Absorption is important because it allows a vapour to be pumped to a pressure far below its saturated vapour pressure, since if the fractional coverage can be kept small, the saturation vapour pressure of the surface can be reduced to a very small fraction of the saturation pressure of the vapour over its own condensate.

3.9 ELECTRICAL CLEAN-UP

It was observed as long ago as 1858, by Plücker, that the pressure in a low pressure discharge tube decreases with time and that the voltage required to establish and maintain the discharge rises with time. In the case of an electrical discharge in a cold-cathode tube, the potential distribution between cathode and anode is not linear, but mostly concentrated in the cathode region (Fig. 3.11).

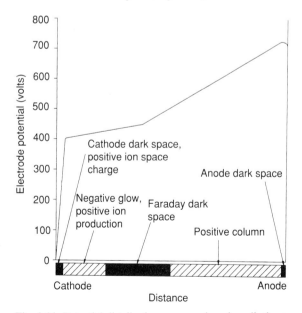

Fig. 3.11 Potential distribution across a d.c. glow discharge.

This cathode drop varies from about 50 to 500 volts depending on the nature of both the electrode and the gas, the gas pressure, the relative area of the cathode and anode and the current. For most of the rest of the potential distribution there is a constant positive potential gradient, whilst at the anode there may be a voltage drop, a voltage rise or no change, depending on operating circumstances. Regardless of actual operating conditions, the anode change is very small compared with that at the cathode. The reason for the large cathode fall is that close to the cathode, current is carried largely by slow moving positive ions, the secondary electrons ejected by ion impact are accelerated away to energies sufficient to initiate ionization and excitation of the background gas species, so that at a small distance from the cathode, ionization starts and the negative glow begins. It is maintained over that distance in which the electrons still have sufficient energy to initiate ionization by inelastic collisions. At the beginning of the Faraday dark space, electrons have energies too small to cause further ionization. This region lasts until the positive column, where ions are formed again; here electrons and ions have equal densities. The anode dark space comprises mainly ions.

In the cathode fall region the positive ions are accelerated through the potential drop V_p and we can write that

$$\frac{1}{2} m_p v_p^2 = V_p e \qquad (3.17)$$

where m_p and v_p are the mass and velocity of the positive ion respectively, e the electron charge and V_p the cathode drop. We can express (3.17) in terms of the equivalent temperature T of the ion thus

$$T = 7.73 \times 10^3 \, V_p \qquad (3.18)$$

The striking fact about this result is that the equivalent temperature of the positive ion is very high.

The ions formed in the vicinity of the cathode are accelerated towards the cathode and strike it with such high velocity that they are driven into the cathode to some extent. There are three consequences of this impact: one is sputtering of cathode material as described in section 3.7, alternatively the ion is simply buried; regardless of this the ion will be neutralized close to the metal surface, with high efficiency, causing an electron to be ejected. During their transition to the cathode the positive ions can collide with neutral molecules and, by a process known as symmetrical charge exchange, produce a fast neutral molecule plus a thermal energy ion

$$M_2^+ + M_2 \to M_2 + M_2^+$$

fast slow fast slow

according to the scheme above. The resulting fast neutral molecule, equivalent in energy to the ion which produced it, but otherwise unchanged, will bombard the cathode in place of the ion, since there is little direction change in the collision; the neutral molecule is driven into the cathode in the same way as an ion would be. In addition to the production of ions and fast neutrals some molecules are dissociated into very reactive atomic fragments.

We may summarize the situation at the cathode surface in a cold cathode discharge as follows:

1. Ions are formed in the negative glow at the edge of the cathode dark space and accelerated towards the cathode by the cathode drop potential; the cathode dark space is filled with positive ion space charge.
2. Ions impacting on the cathode cause sputtering and ion burial.
3. Ions impacting on the cathode are efficiently (~99%) neutralized yielding ejected electrons (secondaries).
4. Some ions are converted to fast neutral species which are capable of sputtering and burial effects.
5. Some molecules are dissociated into extremely reactive atomic species.

In hot cathode devices, e.g. ion gauges, the situation is different since now electron supply is by thermionic emission rather than ion neutralization. The value of the current passing from the cathode to the anode is limited only by 'space charge'. When the gas pressure in such a device is raised, collisions with gas molecules occur, and if the accelerating potential exceeds the ionization potential of the gas, positive ions are produced. For every positive ion a further electron is released so that if the pressure is raised substantially the electron current rises sharply, accompanied by radiation in the visible region (blue glow). Because there is now no substantial cathode drop to accelerate the ions, cathode sputtering is negligible, and furthermore, the hot cathode discourages adsorption of ions and excited species.

The artificial removal of gas species during the operation of a cold cathode or hot cathode device can lead to incorrect measures of the main chamber pressure if these devices are pressure gauges, such as the Penning gauge or Bayard–Alpert gauge. The gauges, in effect, exhibit pumping action. This pumping action can occur even if the gauge envelope is glass rather than metal since a barely discernable film of metal will form on a glass gauge envelope during operation as a result of electrode evaporation or sputtering in normal use. The effective pumping speed of most gauges is very low so that the effect is usually measurable only at very low pressures (UHV).

3.10 ELECTRON AND ION STIMULATED DESORPTION

Electrons and ions incident on solid surfaces can release significant amounts of adsorbed gases. At low pressures (UHV), these processes may well limit the attainable vacuum. When a chemisorbed layer on a metal is bombarded with a beam of electrons, part of their kinetic energy is expended in excitation of the adspecies. Since the impact time is much less than the period of vibration of the chemisorption bond, excitation of the adspecies occurs by a Franck–Condon process and a vertical or energetic transition results. The probability of such excitation is greatest at the maximum displacement of the vibrating adspecies at all vibrational levels except the ground state. Figure 3.12

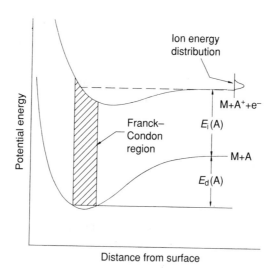

Fig. 3.12 Schematic potential curves for the interaction between a surface M and an atom A, and between M and the ion A^+ (the upper curve): this curve simply represents the state $M+A^++e^-$, and is separated from the ground state curve by the ionization potential of the free atom $E_i(A)$. Following electron bombardment, transitions may occur from the bonding M+A curve to the repulsive part of the upper curve, and ions may desorb with energies as indicated in figure inset.

shows the potential energy curves involved in such a process. The
output of the desorption process is a mixture of neutrals and adsorbate
ions. Desorption of ions results from electronic excitation to an
antibonding state.

The ionization of an adspecies is analogous to gas phase ionization
by electron impact. Accordingly, we define the probability of effecting
ionization, as the ionization cross section, with units of cm^2. The value
for adspecies is of the order of 10^{-17}–10^{-22} cm^2, whilst for the gas phase
process it is much larger at 10^{-15}–10^{-16} cm^2. The flux of atoms or ions
emerging from the surface, as a consequence of this desorption process
can be as high as 10^{-2} per electron. The spatial distribution of the
desorbed ions reflects the nature and geometry of the adatom/surface
bonding. Gas release by this method is capable of causing serious
errors in pressure measurements with the Bayard–Alpert gauge or in
residual gas analysers. Ion stimulated desorption occurs by a process
akin to sputtering, in which adatoms are removed by direct knock-on
collisions with noble gas ions. This process is responsible for gas
release in sputtering systems, Bayard–Alpert gauges and glow
discharges.

There is a further possibility in the presence of electron and ion
impact and that is the initiation of chemical reactions. This is
sometimes used to effect surface clean-up of carbon. For example, in
the presence of hydrogen or oxygen, methane or carbon monoxide can
be formed. In the semiconductor industry, in particular, these effects
are used deliberately to enhance etching in plasma etching processes.

3.11 DIFFUSION

Diffusion has been touched upon in section 1.15. What we must now
consider is a somewhat different process whereby gas is transported
through a solid rather than another gas. In this case, gas is diffusing
through the walls of a vacuum container and contributing to the
outgassing of the system.

The driving force for diffusion is always a concentration gradient.
At the surface of a solid exposed to the vacuum, gas molecules or
atoms within the solid experience a concentration gradient which drives
them towards the surface where they desorb. The rate-limiting step in
this situation is diffusion, which is much slower than desorption. The
diffusion process is described by Fick's law

$$\frac{dC}{dt} = -D\frac{dC}{dx} \qquad (3.19)$$

which relates the flux dC/dt, to the concentration gradient dC/dx through the diffusion coefficient D. Solutions to Fick's law exist for a variety of boundary conditions. The outgassing rate (quantity per unit area per unit time) from a thick solid wall containing gas at an initial, uniform concentration C_0 is given by

$$q = C_0 \left(\frac{D}{t} \right)^{1/2} \left[1 + 2 \sum_{n=1}^{\infty} (-1)^n \exp \left(-\frac{n^2 x^2}{Dt} \right) \right] \qquad (3.20)$$

where D is the diffusion coefficient, x the thickness of the material, n an integer and t the time. For small times, $q \sim C_0(D/t)^{1/2}$ and q varies with $t^{-1/2}$. For long times, the exponential terms must be included and q will vary with $t^{-1/2}\exp(-K/Dt)$ where K is a constant dependent on temperature.

Now the diffusion coefficient D is itself a sensitive function of temperature, according to the relationship

$$D = D_0 \exp(-E_d/k_B T) \qquad (3.21)$$

where D_0 is the diffusion constant and E_d is the activation energy for the diffusion process. Both D and D_0 have the units $m^2 s^{-1}$. A consequence of the exponential dependence on temperature is that a small increase in temperature produces a large increase in the out-diffusion rate, with a consequent reduction in the time required to deplete the total quantity of gas contained in the solid. This effect shows that baking of the surfaces in a vacuum system will have a very advantageous effect in lowering the final pressure attainable or, alternatively, in reducing the time required to reach a specified pressure.

3.12 PERMEATION

Permeation is a three-step process. First, gas adsorbs on the outer wall of a vacuum vessel, then it diffuses through the bulk and finally desorbs from the interior wall. If equilibrium has been established, gas will desorb from the vacuum chamber wall at a steady rate and the process of permeation then behaves exactly like a small leak. The permeability of the vacuum chamber wall for a gas is given by

$$K_p = DS' \qquad (3.22)$$

where D is the diffusion constant of (3.17) and S' the solid solubility in m^3 per m^3 of metal. Consequently, permeability has an exponential dependence on temperature and increases rapidly with temperature. Since S' is dimensionless, K_p has the same dimensions as the diffusion coefficient and shows a similar dependence on temperature. We can write that the steady state permeation rate for a non-dissociating gas is given by

$$q_k = \frac{K_p p}{x} \qquad (3.23)$$

where q_k is the total flux of gas in units of Pa ms^{-1} or Wm^{-2}, and p is the pressure drop across a wall of thickness x. Permeability K_p has units of m^{2}s^{-1} and is really the quantity of gas in m^3 at STP flowing through 1 m^2 of material that is 1 m thick, where the pressure difference is 1 atmosphere. Diatomic gases will dissociate in the first step of the permeation process, namely adsorption, and will then diffuse as atoms. Equation (3.23) is then modified to yield

$$q_k = \frac{K_p \left(p_2^{1/2} - p_1^{1/2} \right)}{x} \qquad (3.24)$$

where p_2 and p_1 represent the pressures on each side of the vacuum vessel wall. Now K_p has the units Pa$^{1/2}$ m^{2}s^{-1}. Equation (3.24) describes permeation through metals since dissociation does not occur on glass or ceramics.

Permeation is, of course, only a serious problem when efforts are being made to produce ultrahigh vacuum (UHV). In particular, hydrogen and helium permeate through glass so that the permeation of atmospheric helium is one of the limiting factors in the attainment of UHV in glass vacuum systems, although glasses do exist (e.g. Monax) which are relatively impermeable to He. Any gas which is soluble in a metal may permeate it since its permeability is directly proportional to its solid solubility. Hydrogen, for example, will permeate most metals and its permeation rate will be described by (3.24) since it dissociates on adsorption. Hydrogen permeation is particularly significant in palladium and this fact is used as the basis of the technique for admitting pure hydrogen to a vacuum system via a heated palladium 'thimble'. Although the permeation of hydrogen through stainless steel is small, it is still sufficiently great to pose a potential problem for ultrahigh vacuum systems. Hydrogen permeation is least in aluminium and alloys such as Duralumin. As might be expected, the permeation rates are highest for the smallest atoms or molecules such as helium or hydrogen, with a decline in permeation rate with increase in molecular diameter.

3.13 OUTGASSING

The preceding sections of this chapter have described the phenomena which may contribute to the gas load in a vacuum system. Clearly, if the gas load is not reduced or removed then the performance of high and ultrahigh vacuum systems will be affected. It is convenient to summarize here the principal elements of this gas load:

1. gas dissolved in the fabric of the vacuum vessel, be it metal or glass;
2. gas which is physisorbed or chemisorbed on the interior surfaces.

Under heading (1) we have, of course, gas which is permeating through the chamber wall having initially adsorbed on the outside surface, as well as gas which dissolved in the material during original fabrication such as melting and casting. It is possible to reduce the effects from this source by careful pretreatments of the material used for fabrication so that vacuum melted steels might be used. Alternatively, materials may be chosen which reduce the mobility of the dissolved gas e.g. using stainless steel wherein the chrome oxide forms a barrier to outgassing. Finally, simply lowering the temperature will immobilize gas inputs via this source.

Gas inputs under the second heading may be dealt with by raising the temperature of the vacuum vessel so that the desorption rate is enhanced. Although the permeation rate is also enhanced, desorption from the surface occurs preferentially, so that when the bakeout temperature is reduced the gas load from desorption is essentially eliminated. For stainless steel systems the usual gases evolved during bakeout at 150–200° C are H_2O, CO_2, CO and H_2, with H_2O and H_2 forming the predominant components of the desorbed gas. Glass systems evolve primarily water during bakeout. For a glass system the bakeout temperature will be much higher, ~ 450° C, reflecting the greater difficulty with which the water is removed. Water may exist on glass in layers up to 50 monolayers thick and although this water is removed by baking, further bakeouts will release more structurally bonded water, i.e. water that is released by diffusion.

We can summarize the basic problems involved in obtaining a vacuum by considering the rate-limiting processes which operate during the pumping of a vacuum chamber. The most obvious step is the first one in which the volume gas is removed by the operation of a suitably chosen pump. Thereafter surface desorption effects are dominant until the surface population is so depleted that diffusion takes over. Finally, only permeation operates, at a constant rate. Figure 6.5 illustrates the time-scale of these processes.

PROBLEMS – CHAPTER 3

3.1 A vacuum system, having a volume of 10 litres, is designed to operate under UHV conditions. If the vacuum system is contaminated by a small amount of *volatile* impurity having a molecular weight of 70, calculate the maximum amount of this impurity which can be tolerated if the threshold for UHV, 10^{-6} Pa is just to be attained at room temperature, 293 K.

3.2 A vacuum system is constructed from stainless steel with a wall thickness of 1 mm and a total surface area of 0.5 m^2. It is pumped by a pump having an observed pumping speed of 100 ls^{-1}. Calculate the lowest pressure obtainable with this arrangement. If the vacuum system were to be constructed from glass but all other parameters remain unchanged, what is the ultimate pressure attainable? Note: the partial pressure of hydrogen in the atmosphere is 0.05 Pa; the permeation constant for hydrogen in stainless steel at 20° C is 4.47×10^{-15} m^2 Pa$^{1/2}$s^{-1}; the partial pressure of helium in the atmosphere is 0.5 Pa; the permeation constant is 2×10^{-13} m^{2}s^{-1} at 20° C.

3.3 The outgassing rate for a particular sample of untreated, unbaked, stainless steel is around 5×10^{-5} Wm^{-2}. It is used to fabricate a vacuum chamber having an internal surface area of 5 m^2. Calculate the pumping speed, in ls^{-1}, which is required to produce a vacuum of 1×10^{-4} Pa. What pumping speed is required if a base pressure of 1×10^{-7} Pa is to be reached?

4

Vacuum pumps; the physical principles

4.1 INTRODUCTION

In order to create a vacuum in a chamber the total gas momentum stored therein must be reduced. This can be done in two distinct ways, either by providing sufficient additional momentum transfer to the gas at some point in the chamber, so as to cause the gas to flow out of the chamber, or by removing the existing momentum, *in situ*, by bonding the gas atoms to a surface; this latter process may be considered entrapment and is reversible for some pump designs.

The first approach provides momentum transfer by interaction with a moving substance, which may be the piston or rotating vane of a mechanical pump, or the working fluid of a diffusion pump. The second approach involves either chemisorption, the controlled chemical interaction with a material introduced for the purpose, for example, an evaporated titanium surface, or alternatively, physisorption, interaction with a heat sink or refrigerated surface. There are momentum removal mechanisms which are hybrids of the two approaches outlined above. Thus in the getter-ion pump, inert gas atoms are given additional momentum from an electric field, but do not flow out of the enclosed space; instead they are 'buried' in a titanium cathode.

We may also distinguish within these two groups those pumps which may be operated by themselves, i.e. those which pump air or gas from a vessel initially at atmospheric pressure, and those which can only begin to operate at a certain pressure considerably less than atmospheric and hence require a fore-vacuum or backing vacuum. Within the first group we find the water-jet pump, the oil-sealed rotary pump, the hook and claws pump and the sorption pump. In the second group are the Roots pump, the turbomolecular pump, the vapour or diffusion pump, the getter-ion pump and the cryopump. The Roots pump can operate without a backing pump, but this is not generally a satisfactory arrangement. The vacuum pump family is summarized in Fig. 4.1.

Usually a vacuum system will comprise a backing pump, generally of the oil-sealed rotary type, coupled to a diffusion pump. The diffusion pump discharges to the rotary pump which in turn discharges to the atmosphere. An alternative arrangement, for those situations where high gas loads are not anticipated, is the use of a sorption pump backing an ion pump.

The introduction and development of essentially all of the important pumping techniques is due to one man, Wolfgang Gaede, who collaborated closely, during his working life, with what was then the small company Leybold's Nachfolger of Cologne, Germany, (now Leybold–Heraeus GmbH). Gaede must be considered a genius of vacuum technology since, beginning in 1907 with the invention of the oil-sealed rotary pump, he followed that with the molecular drag pump (1912) and the diffusion pump (1915). He also, in 1935, introduced the gas ballast concept for oil-sealed rotary pumps. All of these pump types, with the exception of the molecular drag pump which has been superseded by the turbomolecular pump, are still extremely important and dominate modern vacuum technology.

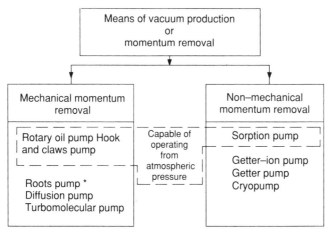

Fig. 4.1 The vacuum pump family, classified according to manner of momentum removal. *The Roots pump can operate from atmospheric pressure, but is not generally used that way.

4.2 TYPES OF MECHANICAL PUMP

The earliest forms of vacuum pump were mechanical pumps and they contributed in large measure to progress in vacuum technology. Amongst these early designs in use at the end of the 19th century, mention must be made of the hand-operated Geissler–Toepler pump, based essentially on the principle established by Torricelli; the operation of this pump is extremely tedious and its pumping speed is low. The water-jet suction pump is still in use in simple laboratory filtration operations, while the Sprengel mercury pump, which operates on the same principle and once rendered valuable service, has now passed into history. No detailed description of these pumps will be made here, since, with the exception of the water-jet pump, they find no use in modern practice.

In 1905, Kaufman introduced a rotary mercury pump which was a great improvement on the previously available pumps such as the Geissler–Toepler pump. Kaufman's design was improved by W. Gaede at about the same time and Gaede's rotary mercury pump soon found acceptance. However, Gaede, the 'father' of modern mechanical pump technology, went on to introduce the rotary oil pump which soon became the most important mechanical pump.

4.3 THE ROTARY OIL PUMP

The modern rotary oil pump is a refinement of the original 1905 design of Gaede. Figure 4.2 shows a section through such a pump. The pump

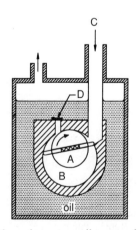

Fig. 4.2 Cross section through a rotary oil pump where A is the rotor, B the stator, C the inlet port and D the exhaust valve.

consists of a steel cylinder, or rotor, A, which rotates eccentrically inside a steel casing, or stator, B, almost touching the stator surface. The rotor A is slotted at its diameter to take two spring-loaded vanes which bear tightly against the inner surfaces of the stator. All of the contacting surfaces are ground to a high precision and are immersed in oil in the pump casing; this oil functions as both a seal and a lubricant.

As the rotor revolves, air enters at the inlet C and is compressed to greater than atmospheric pressure before being expelled through the exhaust valve D. This happens twice in every revolution, as the vanes come past the exhaust valve and, as a consequence, we can say that the free air displacement of the pump, S_D, which is its speed at atmospheric pressure, is given by

$$S_D = 2fV \qquad (4.1)$$

In this equation, f is the number of revolutions per minute of the rotor and V is the maximum volume entrained between the two vanes. If V is in litres then the free air displacement will be in litres min^{-1}.

The pumping speed is quoted at the pressure prevailing at the inlet, so it would be expected that this speed would be constant regardless of the pressure because V is constant. However, as lower pressures are obtained there is, in effect, a leak of gas into the pump. This arises from the small, but vital, clearance between the rotor and stator, the small minimum swept volume between the rotor and stator occurring when the vanes have just passed the exhaust valve aperture, the vapour

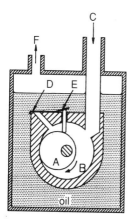

Fig. 4.3 Cross section through a single vane pump of the Cenco–Hyvac type where A is the rotor; B is the stator; C is the inlet port; D is the exhaust valve; E is the vane spring and F is the exhaust port.

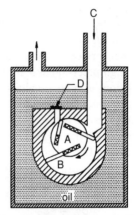

Fig. 4.4 Cross section through a triple vane pump. Here, A is the rotor carrying three spring-loaded vanes which bear on the stator surface B; C is the inlet port and D is the exhaust valve.

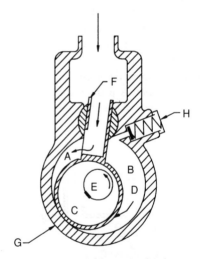

Fig. 4.5 Cross section through a rotating eccentric piston pump of the Kinney type.

pressure of the oil and the fact that the oil itself tends to decompose into gases, particularly at localized hot spots resulting from frictional drag. In modern practice this design has seen only five changes, namely, the single-vane rotary pump, Fig. 4.3; the triple-vane pump, Fig. 4.4; the rotating eccentric piston pump, Fig. 4.5; the multistage

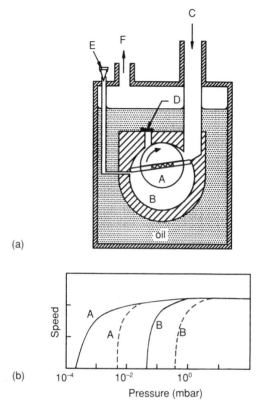

(a)

(b)

Fig. 4.6 (a) Section through a rotary oil pump with gas ballast where A is the rotor; B is the stator; C is the inlet port; D is the exhaust valve; E is the gas ballast air inlet valve and F is the exhaust port. (b) Pumping speed versus pressure characteristics typical of rotary oil pumps. The curves labelled A are for a two-stage pump. The curves labelled B are for a single stage pump. The solid lines are the performance without gas ballast; the broken lines indicate the performance with gas ballast.

pump, to give higher pumping speeds, and the gas ballast pump to deal with soluble or condensable vapours, Fig. 4.6.

The single vane rotary pump, an American design often found as the 'Cenco–Hyvac' design, and the triple-vane pump have the same mode of operation as the two-vane type already described. The rotating eccentric piston pump is, however, significantly different in operation and is often used to extend the speed range of mechanical pumps, since the rotary vane pump seldom exceeds 7000 lmin⁻¹.

The rotating eccentric piston pump, Fig. 4.5, is another American design, this time due to the Kinney Manufacturing Co. It does not use vanes at all; instead it has a tube of rectangular cross section, F, which is a sliding fit in a small auxiliary cylinder and connects the intake port to the rotor. The rotor is mounted eccentrically about the motor driven axle E and is in two parts; the inner cylinder C is keyed to the axle E and rotates with it, but the cylindrical shell D is a sliding fit on C and does not rotate with it because it is rigidly attached to the sliding inlet tube F. The point of contact between D and the stator casing G sweeps round the inner wall of the stator.

As the rotor moves in the direction of the arrow, the volume associated with the space A increases and the corresponding volume associated with the space B decreases. When the eccentric piston has reached its highest point, it expels all air or gas plus surplus sealing oil through the exhaust valve H. Further movement of the eccentric piston enlarges the space A to its maximum volume and cuts it off from the inlet valve. This trapped gas is now compressed and expelled during the next revolution.

The multistage pump simply comprises two conventional vane pumps in series mounted on a common shaft and needs no further comment. The gas ballast pump was introduced by Gaede as a means of reducing the extent of vapour contamination of the oil during the compression cycle. The ultimate pressure attainable by a rotary pump is of the order of 10^{-1}–10^{-2} mbar for a single stage pump and 10^{-3}–10^{-4} mbar for a double stage pump. This pressure is limited by leakage between the high and low pressure sides of the pump, due mainly to carry-over of gases and vapours dissolved or condensed in the oil, which evaporate when exposed to the low inlet pressure, as has already been pointed out.

The gas ballast system reduces the extent of vapour contamination of the oil during the compression stages of pumping, particularly in the presence of water vapour which is a usual constituent of a vacuum system. When water is present in a vacuum system, it turns out that if the compression ratio of the pump exceeds approximately 8:1, water will condense. The resulting liquid will in part be expelled with the gas, but some will emulsify with the oil and evaporate again into the chamber connected to the inlet port: the total pressure will be significantly increased. Now, the higher the compression ratio the better from the point of view of creating a vacuum; for example, with an intake pressure of 1 mbar the compression ratio must exceed 10^3:1 in order to discharge gas to the atmosphere, and compression ratios up to 10^6:1 are found.

The solution proposed by Gaede, Fig. 4.6(a), was that atmospheric air be admitted to the pump during the compression cycle to reduce the effective compression ratio and thereby increase the proportion of non-condensable gas in the pump. By this means, the partial pressure of the vapour being pumped does not exceed its saturation vapour pressure at the time the exhaust valve lifts (the exhaust valve lifts earlier in the pump cycle than it otherwise would) and consequently vapour is discharged without condensing. The extra work entailed in compressing the gas introduced as gas ballast causes a temperature rise which also assists in preventing vapour condensing within the pump.

The operation of gas ballast reduces the compression ratio of the pump to around 10:1 or less and in operation will be potentially capable of removing vapours of up to 40 mbar partial pressure. It also has the effect of transporting oil from the pump chamber and this oil appears as an oil mist. Since gas ballasting will usually be conducted for 20–30 minutes at a time, it is necessary to monitor the pump oil level. The reduction in compression ratio accompanying gas ballasting causes a reduction in the ultimate pressure attainable, usually by around two orders of magnitude. In a two-stage pump, only the final stage is gas ballasted so that the final pressure is not so adversely affected, Fig. 4.6(b).

It need hardly be mentioned that the pump oil is usually a hydrocarbon oil chosen for its low vapour pressure. The oil must also possess the appropriate viscosity for the pump, since too low a viscosity will result in noisy pump operation and too high a viscosity may result in seal failure, loss of vacuum and possibly pump seizure. The nature of the pump oil may be altered to cope with particular pumping situations, thus detergents and inhibitors may be added to the oil, the first named to ensure that any contamination is uniformly distributed throughout the oil, whilst the inhibitors protect the metal pump surfaces from corrosion. There is a penalty for this type of oil: the ultimate pressure is higher than would otherwise be the case and the oil is slightly hygroscopic, necessitating the use of gas ballast as a routine measure. If it is necessary to pump reactive gases which would combine with a simple hydrocarbon, then perfluoropolyether fluids (Fomblin) must be used. These materials are inert to most reactive chemicals, such as O_2, O_3, F_2, UF_6, and they do not polymerize when exposed to electron and ion bombardment. Fomblin is also inactive to amines, does not hydrolyse to form hydrochloric acid (as do the fluoro-chloro oils) and is insoluble in most inorganic solvents. The general inertness of the Fomblin oils leads to reduced pump maintenance and longer operational life.

The oil-sealed rotary pump is the usual choice as a fore-pump, that is, the pump providing the starting vacuum for pumps embodying different operational principles, for example, the diffusion pump or turbo-molecular pump. It is available in a huge range of pumping speeds typically ranging from $45\,l\,min^{-1}$ to around $7800\,l\,min^{-1}$. The ultimate pressure attainable using these pumps can be as low as 1.3×10^{-4} mbar.

4.4 HOOK AND CLAWS PUMP

The hook and claws pump is a recent design introduced by Leybold of Germany. The pumping principle is explained in Fig. 4.7, which shows one set of contrarotating, barely contacting, hook and claws which effect dry compression of gas taken in from the vacuum chamber at the inlet slot and discharge the compressed gas through the exit slot set below the plane of the intake slot. Gas is thus pumped vertically, i.e. in a line at right angles to the plane of Fig. 4.7. Usually, four identical stages of hooks and claws are mounted, one above the other and pinned

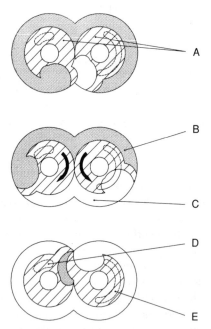

Fig. 4.7 Pumping action of the hook and claws pump where A is the contrarotating hook and claw rotors; B is the compressed gas; C is the inlet space; D is the exhaust slot and E is the intake slot.

to a common drive shaft: they thus operate in series. The pump chambers are kept free of all sealants and lubricants while the end bearings are lubricated with perfluoropolyether (PFPE) grease.

The hook and claws pump may be used to pump condensable vapours or even small amounts of liquid, but its principal application is in the semiconductor industry. Pumping speeds up to $100\,ls^{-1}$ with ultimate vacua of 4×10^{-3} mbar are attainable. Higher pumping speeds may be achieved by combining the hook and claws pump with the Roots-type pump.

4.5 ROOTS-TYPE PUMP

The Roots-type pump was introduced in 1857 by E. Root and combines small physical size with a pumping speed range which is most efficient in the interval between rotary pumps and vapour diffusion pumps, i.e. they have high pumping speed in the range 10^{-3}–10^{-4} mbar. It utilises two figure-eight-shaped rotors which, synchronized by external gears, counterrotate in a chamber **without touching each other or the chamber wall**. The accurate shape of the rotors together with close operating tolerances allows the rotation process to proceed without friction, without lubrication and without substantial gas backflow at pressures below 30 mbar. The rotational speeds are between 500 and 3000 rpm depending on the size of the pump. Figure 4.8 is an operational diagram of the Roots pump.

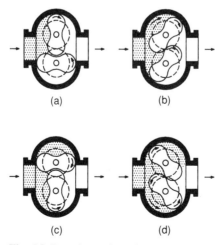

(a)   (b)

(c)   (d)

Fig. 4.8 Pumping action of the Roots pump.

The air or other gas in the chamber to be exhausted enters the Roots pump on the inlet side, Fig. 4.8(a). As the rotors advance, this gas flows into the area shaded in Fig. 4.8(b) until a point is reached where a definite volume of air is trapped, Fig. 4.8(c). Further rotation, Fig. 4.8(d), causes this trapped volume of air to be connected to the outlet or fore-pressure side of the pump and discharged. For each complete revolution of the drive shaft, this process is carried out four times.

Although the Roots pump can discharge directly to the atmosphere without backing pumps, this is not usually desirable. The addition of a backing, for example, a rotary or a hook and claws pump, makes it possible to operate the Roots pump at lower inlet pressures and simul-taneously reduces the power required to drive the pump. The speed and operating pressure range of the pump depend in part upon the perform-ance characteristics of the backing pump used. A low ultimate pressure for the backing pump results in a low ultimate pressure for the com-bination. The higher the speed of the backing pump the higher the speed of the combination. A balance between speed and ultimate pres-sure can therefore be achieved by suitable choice of backing pump. Thus, it is possible to reach ultimate pressures of less than 5×10^{-4} mbar, with a peak pumping speed of $5200 \, \mathrm{ls}^{-2}$ using a Roots pump backed by a two-stage rotary pump.

4.6 THE MOLECULAR DRAG PUMP; TURBOMOLECULAR PUMPS

The molecular drag pump, first described by W. Gaede in 1912, is very different in character from the pumps described thus far in as much as **there is no mechanical separation between the high and low vacua**. It depends for its operation on the laws of gas flow at very low pressures, determined by Knudsen, Smoluchowski and also Gaede himself.

The fundamental principle of the pump may be illustrated by means of the schematic diagram, Fig. 4.9. The cylinder A rotates on its axis in the direction indicated by the arrow, within the airtight stator B, and **drags** gas from the inlet x to the outlet y. This results in a pressure difference between x and y as demonstrated by the manometer. Struc-turally, there is a slot in the case between x and y whilst everywhere else the stator is a close fit on the rotor.

At ordinary pressures, as pointed out in Chapter 1, viscosity is independent of pressure and the pressure difference between x and y

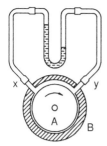

Fig. 4.9 Principle of operation of the molecular drag pump.

depends only on the angular velocity of rotation, ω, the gas viscosity η, the length ℓ of the slot and its height h measured radially. The relationship between these quantities was shown by Gaede to be given by

$$p_2 - p_1 = 6\frac{\ell\omega\eta}{h^2} \tag{4.2}$$

where p_2 is the pressure on the outlet side, y.

At low pressures, or when the mean free path is comparable to the characteristic dimension, the number of intermolecular collisions becomes very small and molecule/surface collisions dominate. Under these conditions, molecules striking a surface take up the direction of motion of the surface and (4.2) is no longer valid since viscosity is no longer independent of pressure, see Chapter 1. Now, instead of the **pressure difference** remaining constant, for constant rotation speed, we have the **pressure ratio constant**, independent of the fore-vacuum pressure.

Gaede showed that, at very low pressures

$$\frac{p_2}{p_1} = \exp(b\omega) \tag{4.3}$$

where b is a constant whose value depends on the gas being pumped and the slot characteristics, and ω is the angular velocity of the rotor. Since gas molecules have speeds of the order of $500\,\mathrm{ms}^{-1}$ at room temperature, the pump can only exert a dragging effect if the speed of the rotor is very high and h is very small. The constant b increases with decreasing h.

Until the end of the 1950s, interest in molecular drag pumps was small, in large measure due to the problems associated with maintaining very small clearances and operating bearings at the high rotational speeds required. A new design introduced by W. Becker in 1958, and built in Germany by Pfieffer GmbH, known as the turbomolecular pump or axial flow molecular turbine, at last removed the problem of rotor seizure by eliminating the need for very small (0.02 mm) clearance between rotor and stator. A section through Becker's design is shown schematically in Fig. 4.10(a). It is important to realise that this pump is not actually a **molecular drag** pump, instead gas molecules acquire directed momentum through collisions with a turbine blade moving at high speed.

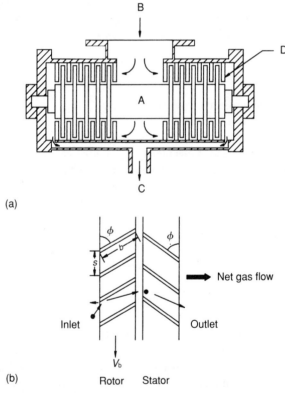

Fig. 4.10 (a) Simplified cross section through a dual-rotor turbomolecular pump. Here, A is the high speed rotor; B is the inlet port; C is the exhaust port and D is the stator blade. (b) Sectional view of a rotor/stator disc blade arrangement.

The rotor is now in the form of two sets of parallel, radially slotted discs. In action these slotted discs rotate between similar radially slotted stationary discs attached to the cylindrical pump housing. An intake port for the pump is provided with direct access to the region between the two sets of discs. The exhaust port combines the output from both sets of blades, which is then removed by a backing pump. The physical arrangement of the slots is depicted in Fig. 4.10(b). A rotor disc has everywhere an adjacent stator disc. The radial slots in the discs are angled and arranged so that the slots in the stator disc are mirror images of those in the rotor discs, i.e. the rotor slots are inclined in the opposite way to those in the adjacent stator discs.

In order to understand the mode of operation consider Fig. 4.10(b). The slots in the rotor disc are angled in the opposite sense to those in the stator and the relative velocity between the alternate slotted rotor blades and the slotted stator blades makes it probable that a gas molecule will be transported from the pump inlet to the pump exhaust. As the rotor moves in the direction of the arrow with velocity v_b, gas molecules acquire directed momentum through collisions with the moving blade. The rotor blades are slotted at an angle ϕ to make the probability of a gas molecule being transmitted from the inlet side to the outlet side greater than in the reverse direction. Similarly for the stator blades which are slotted in the reverse sense, and from the standpoint of the molecule which has received directed momentum from the rotor, appear to be moving at high relative speed. The molecule then receives further directed momentum, in the same sense, from the stator. This process occurs at all slots on each disc. With discs a few mm thick, only a short channel is provided between discs and consequently only small pressure differences are attained. However, with several discs, each with many slots all effectively acting in series, a large pressure difference can be established between input and output. With a small pressure difference between one pair of discs, the clearances can be as much as 1 mm without impairing pump performance. This in turn eliminates the problems associated with earlier designs where rotor seizure was commonplace. If the angle between the slot and the surface of the disc is decreased, a higher compression ratio is achieved, but at the expense of a smaller pumping speed. In practice the slot angles are varied along the rotor shaft, thus the discs near the centre or high vacuum port have large slot-to-surface angles to ensure high pumping speed. The outer discs, operating at higher pressure, have smaller angles.

If one considers a single blade row, such as the rotor row in Fig. 4.10(b), then we can write, following the approach of Kruger and Shapiro (1961), that the maximum compression ratio K_{max} of the row is exponentially dependent on the blade speed ratio and some function of the blade angle $f(\phi)$, thus

$$K_{max} \simeq \exp\left[\frac{v_b M^{1/2}}{(2k_B N_A T^{1/2})}\right] f(\phi) \qquad (4.4)$$

where the contents of the bracket represent the blade speed ratio, that is, the ratio of the blade speed to the most probable molecular speed. This result is only true for blade speed ratios ≤ 1.5.

Since the compression ratio is exponentially dependent upon rotor speed and $M^{1/2}$, we note that the light gases such as hydrogen and helium will have compression ratios much smaller than the heavy gases. The compression ratios multiply as the number of stages is increased, so that for the pump represented in Fig. 4.10(a) which has 6 stators and 5 rotors, i.e. 11 stages, we must raise the compression ratio by the power 11 to arrive at the overall value. (A stator blade has the same compression ratio as a rotor blade.) Of course the linear blade velocity is proportional to the radius as well as the rotor angular frequency so that an area closer to the centre of the rotor will have a smaller speed ratio and smaller ratio s/b, the blade spacing to chord ratio, see Fig. 4.10(b). The net effect of this is that the compression ratio becomes smaller as one moves closer to the rotor axis.

The compression ratio of a typical turbomolecular pump may well be 10^8–10^9 for nitrogen while for hydrogen it may be only around 10^3, a consequence of the dependence on $M^{1/2}$. The pumping speed for the two gases may well not differ significantly since it turns out that the net pumping speed of the blade is independent of M and is in fact slightly greater for light gases than heavy ones. The maximum compression ratio occurs only when the pump is pumping no gas, e.g. at its ultimate pressure, whilst the maximum speed occurs, by definition, when the pressure drop is zero. An operating pump works in the range between these two extremes. The ultimate pressure of a turbomolecular pump is actually determined by the compression ratio for light gases. The hydrogen compression ratio in a turbomolecular pump is low enough to cause the ultimate hydrogen partial pressure to be determined, in effect, by the hydrogen pressure in the foreline. Expressed slightly differently, we can expect hydrogen to back-diffuse and be the major constituent of the residual gases. This is born out in practice; the ultimate pressure is

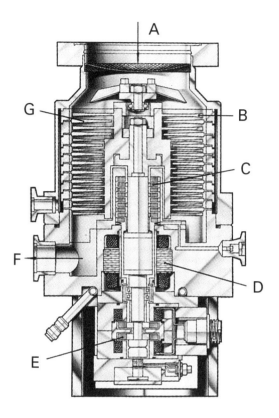

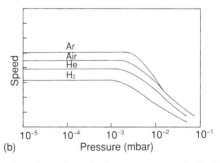

(b)

Fig. 4.11 (a) Section through a single rotor magnetically levitated turbo-molecular pump. This example is a Leybold Turbovac 340 M®, with permission. Here, A is the vacuum port; B is the rotor; C is the permanent magnet bearing; D is the d.c. motor; E is the stabilizer; F is the fore-vacuum port and G is the stator. (b) Typical pumping speed versus pressure curve for a turbomolecular pump.

around 10^{-11} mbar. The turbomolecular pump, however, does also suffer from oil contamination and decomposition since oil is required for the bearings. However, in a more recent design introduced by Leybold AG, frictional contact in the bearings and the consequent need for lubrication has been eliminated. This has been achieved by magnetic levitation of the rotor, Fig. 4.11(a). Pumps based on this approach are entirely free of hydrocarbon contamination and may also be operated in any orientation: older designs frequently had an orientation restriction resulting from bearing design limitations. The pumping speed versus pressure dependence of a typical turbomolecular pump is sketched in Fig. 4.11(b).

Pumps based on the turbomolecular design have special attractions where a particularly clean vacuum is required: they are available up to pumping speeds of 1950 $1min^{-1}$ for air, backed by a two-stage rotary oil pump. They are, though, very susceptible to damage from ingress of solid objects.

4.7 DIFFUSION PUMPS

The vapour diffusion pump is a monument to two men, W. Gaede and I. Langmuir, who developed different versions of the pump at around the same time, Gaede (1915) working in Germany and Langmuir (1916) working in America. In the diffusion pump, the basic pumping action is that a stream of the pumping fluid vapour emerges from an annular nozzle and expands into the pump casing, initially evacuated to a backing pressure of around 10^{-2} mbar, where it condenses on the chilled walls. This vapour, streaming in the direction of the pump discharge outlet at the backing pressure, imparts momentum in this direction to the gas molecules, so creating a pumping action and thereby a pressure gradient in the gas between the high vacuum inlet and the discharge outlet. Further, the vapour stream creates a 'seal' across the mouth of the pump thus preventing gas molecules back-diffusing towards the intake aperture. The physical arrangement is shown schematically in Fig. 4.12.

In order to achieve the effects described above, the velocity of the molecules in the vapour stream must meet certain minimum requirements. What these are can be seen very easily by remembering that the average velocity of a molecule in the pump mouth will be very high, so that the velocity of the vapour stream must be considerably higher if directed motion is to occur.

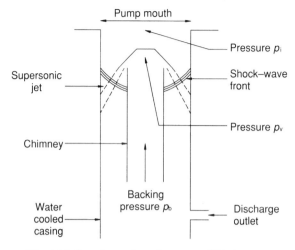

Fig. 4.12 Pumping action of a vapour diffusion pump.

The average velocity $\bar{c}$ of a molecule at absolute temperature T is given by (1.2) as

$$\bar{c} = \left(\frac{8k_{B}T}{\pi m}\right)^{1/2}$$

Now this can be recast in terms of the gas density, ρ, and the pressure, p, as

$$\bar{c} = \left(\frac{8p}{\pi\rho}\right)^{1/2} \tag{4.5}$$

The speed of sound c_S in a gas or vapour is given by

$$c_{S} = \left(\frac{\gamma p}{\rho}\right)^{1/2} \tag{4.6}$$

where $\gamma = C_P/C_V$ and C_P, C_V are the molar specific heats at constant pressure and volume respectively.

We can combine (4.5) and (4.6) to express the ratio between the average molecular velocity and the speed of sound in a gas or vapour, thus

$$\frac{\bar{c}}{c_S} = \left(\frac{8}{\gamma\pi}\right)^{1/2} \tag{4.7}$$

For a diatomic gas, $\gamma = 1.4$; for mercury, a monatomic vapour and one of the earliest pumping vapours, $\gamma = 1.67$, and for the complex modern pumping fluids with many degrees of freedom, $\gamma \simeq 1$. In all cases, $\bar{c}/c_S > 1$ and the vapour stream must be supersonic. For the modern pumping fluid, it must be substantially so, since $\bar{c}/c_S = 1.6$.

To ensure a satisfactory vapour stream, the pump fluid in the pump boiler has to be heated until the pressure, p_v, of its vapour is higher than both the backing pressure, p_b, provided in the pump, and the gas pressure, p_i, at the inlet port. The vapour stream emerging from the nozzle expands adiabatically into the region of lower gas pressure around the nozzle; it will achieve supersonic velocities if the following criterion is met, namely,

$$\frac{p_c}{p_v} = \left(\frac{2}{\gamma+1}\right)^{\gamma/(\gamma-1)} \tag{4.8}$$

In this equation p_c is the critical back pressure, i.e. the pressure outside the nozzle required to ensure supersonic flow, and p_v is the fluid vapour pressure before expanding through the nozzle. For most vapour pump fluids the ratio p_c/p_v is between 0.48 and 0.58.

The supersonic stream of pump fluid vapour travels through the gas in the pump, the vapour molecules collide with, and entrain, gas molecules in the neighbourhood of the pump nozzle. These gas molecules are given a marked velocity component directed towards the pump outlet and backing pump; this establishes a pressure gradient and pumping action. The supersonic, high pressure region of vapour overtakes the slower moving gas molecules which have sonic speeds. There is a consequent pressure rise resulting in a steep and stable wave front, i.e. a shock wave is formed in which the gas is rapidly compressed. This shock wave acts as a 'dam' or 'seal' across the pump so that gas from the backing region cannot surmount the pressure step of the shock wave and return to the high vacuum inlet. If the backing pressure is too high, the shock wave front will be too near the nozzle outlet giving a less satisfactory sealing effect. Gas molecules then back-diffuse to the pump intake aperture, increasing the ultimate pressure. A vapour diffusion pump therefore has a **critical backing pressure**, above which there is a more or less sudden increase of pressure on the high vacuum side. This critical backing pressure is lower than the critical back pressure p_c and depends to some extent on the choice of pump fluid. It lies in the range from 1.3 mbar to 0.4 mbar, according to pump design and pump fluid. Since the vapour diffusion pump operates by gas molecules diffusing into the supersonic pump fluid stream, high pumping speeds can only be achieved by using large intake throat areas, with high conductance: the reader is referred to Chapter 2. Expressed differently, this implies a large inlet diameter and a small nozzle. However, if the nozzle is made too small the supersonic jet stream does not spread out to the water-cooled walls and the seal effect is lost; back diffusion then becomes significant.

Usually the restrictions on vapour diffusion pump design are met by means of a multistage vapour pump, Fig. 4.13(a), where two, three or

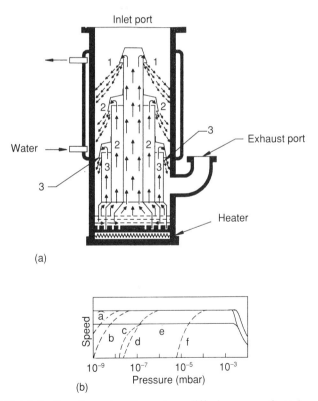

Fig. 4.13 (a) Section through a three-stage diffusion pump featuring self-fractionating. (b) Typical speed versus pressure curves for a multistage diffusion pump, as a function of pump fluid: (a) Santovac 5, (b) Silicone 705, (c) Silicone 704, (d) Apiezon C, (e) Fomblin®, (f) Silicone 702.

even four stages are used, the second stage backing the first and the third backing the second. This multistage design is frequently combined with means to purify the pump fluid; self-fractionating pumps, Fig. 4.13(a). With this multistage design, higher critical backing pressures are permitted. The fractionating effect is achieved by arranging that each pumping stage is fed from its own cylindrical chimney whereas a single chimney supplies all stages in the non-fractionating pump. Purification is achieved by the condensate from all three nozzles entering first the outer annular section of the boiler (3), from which the most volatile fractions of highest vapour pressure boil off to feed the third stage which in turn backs the second stage (2). The first stage (1) is fed from the boiler central compartment and thus receives the least volatile constituents of the oil.

Vapour diffusion pumps are generally operated using a very stable, high-purity hydrocarbon (mineral-based oil) as the pumping fluid, although for particular applications, where hydrocarbons or other organic materials cannot be tolerated, mercury is used. Table 4.1 summarizes the principal pumping fluids in use today. Mercury has special advantages as a pump fluid in that it cannot decompose, however its very high vapour pressure at 20° C means that it can never be used

Table 4.1 Properties of some common diffusion pump fluids

Fluid	Composition	Mol. wt	Ultimate pressure* @ 20° C mbar
Apiezon A	Mixture of hydrocarbons	354	6.5×10^{-5}
Apiezon B	" "	420	1.3×10^{-6}
Apiezon C	" "	479	1.3×10^{-7}
Edwards L9	Naphthalene based	407	5×10^{-9}
Silicone DC 702	Mixture of polysiloxanes	530	6.5×10^{-6}
Silicone DC 703	" "	570	6.5×10^{-6}
Silicone DC 704	Single molecule siloxane	484	6.5×10^{-8}
Silicone DC 705	" "	546	1.3×10^{-9}
Santovac 5	Polyphenylether	446	1.3×10^{-9}
Fomblin 18/8	Perfluoropolyether	2650	2.7×10^{-8}
Mercury	–	201	1.2×10^{-3}

* The ultimate pressure attainable at 20° C is generally an order of magnitude or so higher than the true vapour pressure of the pump fluid at that temperature. The vacuum attainable in a properly cold-trapped system will, however, always be less than this figure, particularly with respect to mercury.

without cold trapping. In practice, however, all pumping fluids are operated with suitable baffles or cold traps to reduce backstreaming, where the pump fluid escapes through the inlet aperture. In many instances this can be reduced significantly by fitting a guard ring round the first stage jet.

The pumping speed for a vapour diffusion pump is constant and limited by the pump mouth diameter, since this determines the number of molecules per second which may enter the pump and become entrained in the supersonic jet. Pumps are available with speeds ranging from around 135 ls^{-1} to 45 000 ls^{-1}. The speed of a vapour diffusion pump is higher for hydrogen or helium because of their low molecular weight; the speed varies inversely with $\sqrt{M}$. The speed for hydrogen or helium may be increased by as much as factor of 50% over the value for air (nitrogen). Speed versus pressure curves do however show some dependence on choice of pumping fluid, Fig. 4.13(b).

In use, the vapour diffusion pump with correctly chosen accessories and appropriate pumping fluid will readily give ultimate pressures of 10^{-9} mbar, and with proper procedures, as described in Chapter 6, this figure can be reduced to 10^{-11} mbar: there is no theoretical limit.

Before leaving the subject of vapour diffusion pumps it is necessary to mention the vapour ejector pump, Fig. 4.14, which, although having a similar working principle, is designed to have a maximum pumping speed at intake pressures of 10^{-2}–10^{-1} mbar and above, operating with backing pressures of up to 1 mbar. The principal distinction between

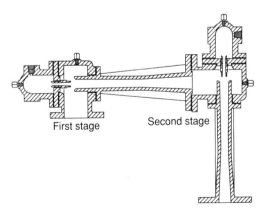

First stage   Second stage

Fig. 4.14 Section through a two-stage vapour ejector pump.

the two types of pump is that, whereas the vapour diffusion pump operates in the molecular flow region, the vapour ejector pump operates in the viscous flow region. The vapour stream now entrains by viscous drag rather than by diffusion; the vapour density in the jet is much higher than it is in a diffusion pump. The usual role of a vapour ejector pump is to operate in the pressure range between 10^{-1} mbar and 10^{-4} mbar, often as an intermediate between a diffusion pump and a rotary oil pump. Its main competitor in this role is the Roots pump already discussed.

4.8 GETTER PUMPS AND GETTER-ION PUMPS

The use of gettering to obtain high vacuum predates the introduction of the vapour diffusion pump in 1915. It offered, in conjunction with a mechanical pump, a simple means of obtaining relatively good vacua. The principles lying behind gettering have been discussed in Chapter 3 and will not be considered further here.

In order to use gettering action in a pump, it is necessary to select a material which offers a high sticking coefficient for most gases and which can be conveniently presented as a very high surface area film. The usual choice for such a material is titanium, although zirconium has been used. Three main types of pump have been introduced utilizing the gettering action of titanium; two are getter-ion pumps and the third is a sublimation, or getter, pump. All of them operate relative to a backing pressure established by either a rotary oil pump or, more commonly, a sorption pump.

The two kinds of getter, or sputter-ion, pump are the diode form, in which a honeycomb anode structure is suspended symmetrically between two titanium plate electrodes, and the triode form in which a stainless steel honeycomb anode is placed symmetrically between two honeycomb titanium cathodes which are, in turn, set in between a stainless steel collector which forms the pump envelope. The pumps operate by sputtering titanium from the titanium cathode in a gas discharge formed by applying $\sim 5\,\mathrm{kV}$ between anode and cathode. A uniform magnetic field of flux density B is established with its lines of force along the common axis of symmetry of the electrodes (Figs 4.15(a), 4.16).

Within this configuration the electrons are confined to spiral paths having axes along, or at small angles to, the axis of symmetry of the electrodes. Consequently, they do not reach the anode immediately, but oscillate backwards and forwards between the cathodes in tight spiral paths before finally being trapped at the anode. As a result, the average

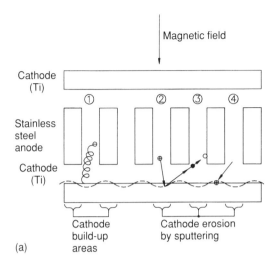

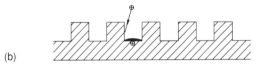

Fig. 4.15 (a) Section through a diode sputter-ion pump showing some pumping mechanisms. (1) The electron field emitted from the cathode initiates the gas ionization process. (2) The ion generated by electron collision bombards the cathode surface causing Ti removal by sputtering. (3) The active gas molecule is chemisorbed by fresh sputtered Ti deposit on the anode surface. (4) Inert gas ions are buried in cathode build-up area and subsequently covered by deposited Ti atoms. Cathode build-up areas occur adjacent to the ends of each anode element. Cathode erosion areas occur opposite ends of each channel. The broken line shows the cathode profile after sustained operation. (b) Channelled cathode for a diode pump giving enhanced sputtering from glancing incidence collision. The sputtered material buries, for example, Ar ions in the bottom of the slot.

electron path is very long and its ability to ionize a gas atom is preserved, even down to pressures of 10^{-10} mbar. The processes involved are the basis of the Penning discharge (the Penning gauge, Chapter 5). The positive ions produced by electron impact are accelerated through the applied potential and strike the titanium cathode. There they are collected and eject not only titanium atoms (sputtering), but previously sorbed gas on the cathode surface (gas sputtering) and also electrons (secondary electrons); these latter electrons join in the ionization

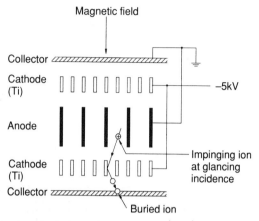

Fig. 4.16 Section through a triode-type getter-ion pump, showing cellular anodes and cathodes.

process. The mass of each positive ion will, on average, be at least 3.7×10^3 times that of an electron and in consequence they reach the cathodes almost directly.

Now, sputtering of titanium from the cathodes results in a fresh, active titanium film being deposited on the anode and it is at the anode that most of the gas will be chemisorbed and held as a titanium compound. For the inert gases the situation is different, since they cannot, in general, form chemical compounds. The inert gases are pumped by first being converted into an ion, accelerated by the electric field to the cathode and then 'buried' in the cathode.

The sputter-ion pump has some useful advantages, namely:

1. There is no pumping fluid, consequently backstreaming does not occur and cold traps and baffles are not required.
2. The pressure attained for a given gas is directly proportional to the discharge current so that, with suitable calibration, the pump current can be used to measure pressure.
3. The pumping speed is roughly constant from 10^{-5} mbar to 10^{-8} mbar.

Against these advantages must be set the following disadvantages:

1. The pump has a long, but finite, lifetime limited by sputtered material flaking off from the anode and short-circuiting the device.

2. Under certain circumstances inert gas (argon) will be released from the cathodes, (cathode memory or argon instability).

3. Hydrocarbons adversely affect pump operation.

Argon instability may be combatted by the introduction of slotted cathodes on diode pumps, Fig. 4.15(b), while the large collector area available on triode pumps means that these pumps exhibit very little memory effect anyway. The slotted cathode and the triode structure are really ways of increasing the argon pumping speed through increase in both surface area and sputtering rate. The increased sputtering rate ensures adequate argon ion burial. More recently, some of the disadvantages outlined above have been reduced in a design which combines the diode and triode structures in one envelope by introducing a secondary plate between the primary cathode and the pump envelope. This pump, the diode/triode pump, is due to Bance (1990).

An alternative approach to pumping, merely using the reactive properties of titanium surfaces, is the getter pump or sublimation pump. Here, titanium is evaporated from a tungsten filament overwound with titanium wire, from a filament of titanium/molybdenum alloy or from a filament of titanium alone, onto a substrate usually at room temperature. Now active gases are pumped by chemisorption: there is no pumping action for inert gases or saturated hydrocarbons, e.g. methane. Usually the evaporation source is operated at constant voltage and the filament current indicates filament life.

Titanium sublimation pumps are free from organic contamination and provide high pumping speeds for active gases. For example, on a water-cooled surface the deposited titanium film will have a pumping speed per unit area, due to gettering action, depending on the sticking coefficient of the gas molecules and the number of molecules incident per unit area per second. With large sublimation rates, pumping speeds up to $10^5 \, \text{ls}^{-1}$ are possible.

We can see how this arises by considering an idealized pump comprising just a blanked-off section of tube, with cross sectional area A, coated on all its inner surfaces with a titanium film. The limiting factor here is just the conductance of the mouth of the pipe for molecular flow. We can use (2.23) to give the conductance (remember pumping speed has the same units), so that for a gas having a sticking coefficient s, the pumping speed S_p of unit area is given by

$$S_p = 3.64 s \left(\frac{T}{M} \right)^{1/2} \qquad (4.9)$$

where S_p is in $ls^{-1}cm^{-2}$. At room temperature, S_p for N_2 and H_2 is $11.8s$ and $44s$ $ls^{-1}cm^{-2}$ respectively. It is easy to see that even for a very modest area of absorbing film, the resulting pumping speed will be very high.

4.9 SORPTION PUMPS

Sorption pumps are physically the simplest pumps available and provide a convenient, clean (oil-free) backing pump for high vacuum pumps such as the getter-ion pump. They have been in use since Dewar first used them in 1905. A sorption pump consists of a cylindrical container with a filling of a material which has a large surface area and sorbs gas by physisorption, as described in Chapter 3. Historically, the first such material was activated charcoal, but nowadays custom-built, activated alumino-silicates (zeolites) are used. They are generally known as molecular sieves. Their structure and properties were considered in Chapter 3.

The essential feature of sorption pump materials is a very high surface area to volume ratio; thus activated charcoal can be prepared with an effective surface area of around 2500 m^2g^{-1}. Although charcoal will absorb gases at room temperature, sorption is greatly increased by cooling to liquid nitrogen temperature, 77 K, and this is a general feature of all sorption pump materials.

A molecular sieve material offers greater sorption capacity at low temperature than activated charcoal and is generally more consistent in its properties, although it does suffer from the fact that its thermal conductivity is considerably less than for charcoal, and it is, therefore, less readily chilled. Molecular sieve materials are provided with specific pore diameters and the popular materials are categorized as 4A, 5A and 13X. These numbers merely categorize the material according to its pore diameter in Ångström units (1 Å – 10^{-10}m), thus 5A has a pore diameter of five Ångströms. Type 5A is the material most commonly used and is calcium alumino-silicate, usually supplied as spherical pellets approximately 3 mm in diameter.

The construction of a sorption pump is simplicity itself since all that is required is a cylindrical stainless steel container, filled with the sieve compound and having an outlet tube of sufficient diameter to give adequate conductance and hence pumping speed. The pumping speed can be, and often is, increased by mounting several sorption pumps in parallel, each immersed in its own Dewar flask containing liquid

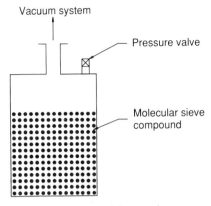

Vacuum system

Pressure valve

Molecular sieve compound

Fig. 4.17 Structure of the sorption pump.

nitrogen, Fig. 4.17. A sorption pump can only pump a limited quantity of gas, determined by its absorptive capacity, after which it will become saturated and begin to desorb gas. To cope with this problem it is always necessary to fit isolation valves which allow the pump to be warmed up, open to the atmosphere. Usually, a safety valve will also be fitted to cope with dangerous over-pressures if an isolated pump happens to lose its liquid nitrogen supply.

The great simplicity of sorption pumps is attended by some disadvantages, namely, they do not pump inert gases, He, Ne or indeed H_2, as well as they pump oxygen and nitrogen; they pump water vapour preferentially and this will often preclude pumping of other gases. As a general rule, molecular sieve material must be baked at ~ 250° C before use, to remove water, since the absorbed water is not released simply by warming to room temperature. Despite these problems, a sorption pump will reduce the pressure in a vacuum chamber to 10^{-2} mbar without any other assistance. The vacuum thus obtained is entirely free of contamination.

4.10 THE CRYOPUMP

The cryopump depends for its operation on physisorption of gases on a chilled surface, as described in Chapter 3. Roughly speaking, the temperature of the chilled surface (cryosurface or cryopanel) determines the ultimate pressure attainable, while the refrigerative power determines the pumping speed. The cryopump, on the face of, it would appear similar to the sorption pump in relying on physisorption on a cold surface for its pumping action. However, there is a significant difference; the temperature of operation of a cryopump is always much lower than that of a sorption pump, usually 4 K. As a consequence it will condense oxygen and nitrogen. Indeed we can assume that all incident gases will have a condensation coefficient of unity.

We can use (1.49) for the molecular collision frequency, to obtain an idea of the pumping speed of a cryopump. Consider a gas of molar mass M impinging on a cold surface of area A, with unit condensation coefficient. The number of molecules N condensing per second is then, from (1.49).

$$N = \frac{pN_A A}{(2\pi MRT)^{1/2}} \tag{4.10}$$

Now N_A molecules occupy a volume V at a pressure p given by $pV = RT$, so that the volume occupied by N molecules at this pressure is NRT/pN_A. If we define the pumping speed S_p of the cryosurface as the volume of gas condensed per second at a pressure p, we note that

$$S_p = NRTA/pN_A$$

so that substituting for N we get

$$S_p = A\left(\frac{RT}{2\pi M}\right)^{1/2} \tag{4.11}$$

or, in simplified form

$$S_p = \frac{62.5A}{M^{1/2}} \ \text{ls}^{-1} \tag{4.12}$$

Equations (4.11) and (4.12) represent the maximum performance attainable. If the condensation coefficients are less than unity, then the speed will be less than the value given by (4.12). It is convenient to consider the situation in terms of the partial pressure p_1 of the condensable gas at the temperature T of the vacuum chamber as a whole, and p_2, the partial pressure at the cryosurface temperature T_c. The number of molecules arriving per unit time at the cooled surface of area A is N_1, given by

$$N_1 = \frac{p_1 N_A A}{(2\pi MRT)^{1/2}} \tag{4.13}$$

The arrival pumping speed S_1, measured at the pressure p_1 is just

$$S_1 = \frac{N_1 RTA}{p_1 N_A} = A\left(\frac{RT}{2\pi M}\right)^{1/2}$$

The number of molecules leaving the cryosurface in unit time is N_2 and

$$N_2 = \frac{p_2 N_A A}{(2\pi MRT)^{1/2}}$$

with a corresponding speed

$$S_2 = \frac{p_2 A}{p_1}\left(\frac{RT}{2\pi M}\right)^{1/2}.$$

The net pumping speed is

$$S_p = S_1 - S_2.$$

Assuming that $T = 293$ K (room temperature) we obtain

$$S_p = \frac{62.5}{M^{1/2}}\left(1 - \frac{p_2}{p_1}\right) \text{ls}^{-1} \tag{4.14}$$

We can usually assume, with some small error, that p_2 is the vapour pressure of the gas at the cryopanel temperature T_c.

The usual choice for a coolant is liquid helium and at liquid helium temperature all gases will solidify except helium itself; hydrogen though will still exert a vapour pressure of $\sim 10^{-7}$ mbar. For very high vacuum some additional means of hydrogen removal is required. The latent heat of vaporization of liquid helium is rather small so that it is always necessary to shield the cryosurface from radiative heat loads. This is generally done by surrounding the cryosurface with liquid nitrogen cooled shielding. Conductive heat loads must also be considered very carefully. These two elements together largely determine the refrigerative power required to sustain the system.

The cryopump is an attractive approach to the provision of very clean vacuum. It cannot sensibly operate from atmospheric pressure

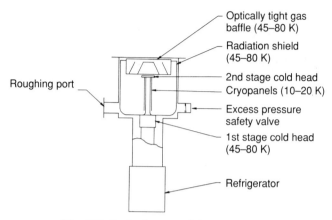

Fig. 4.18 Basic elements of the cryopump.

alone, so it requires some sort of backing pump. The basic design features of a simple cryopump are shown Fig. 4.18. Very high pumping speeds are, in principle, available from quite modest cryosurface areas. Pumps with speeds up to 3500 ls^{-1} are available commercially. In fact, for a given pump mouth diameter, the cryopump offers the highest pumping speed of all high-vacuum pumps with total cleanness. It cannot, however, pump helium to any great extent and, of course, as a gas entrapment pump it has a limited capacity. The capacity of a cryopump for a given gas is just the quantity of gas (in pV units) at $T = 293$ K that can be bound by its cryopanels before the pumping speed for the gas in question drops to 50% of its initial value.

PROBLEMS – CHAPTER 4

4.1 A fifteen stage turbomolecular pump, with a blade tip velocity of 500 ms^{-1}, has a compression ratio for N_2 at 25° C of 7.7×10^8. What is the compression ratio for this pump when it is pumping hydrogen?

4.2 A diffusion pump has a pumping speed of 300 ls^{-1} at 10^{-2} mbar. It has a critical backing pressure of 0.6 mbar. Calculate the maximum throughput and hence the minimum backing pump speed which can be employed for this diffusion pump.

4.3 A circular plate of radius 10 cm and cooled to liquid helium temperature (4 K) is used as an additional pumping element in a diffusion pumped chamber. What is the maximum additional pumping speed which can be obtained from this arrangement for (a) nitrogen and (b) hydrogen?

5

Pressure measurement

5.1 INTRODUCTION; THE VACUUM SPECTRUM

Pressure measuring gauges are now available which, when selected appropriately, enable pressure measurements to be made over eighteen decades. Within this range of pressure measuring gauges we must distinguish two distinct classes; total pressure gauges and partial pressure gauges. The latter are, of course, really mass spectrometers; they offer the particular advantages of high sensitivity together with identification of the constituent gases. We shall in this chapter be concerned with gauges operating mainly in the range 10^3–10^{-15} mbar. This pressure range more or less covers the 'vacuum spectrum', which is itself subdivided into sectors, each usually spanning three decades of pressure. Figure 5.1 shows how these sectors are named.

Table 5.1 indicates the types of pressure gauge which will be considered in this chapter, together with their range of operation on the same pressure scale. In most cases, more than one model of a particular

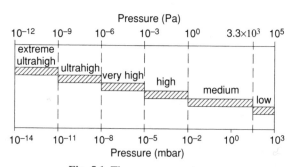

Fig. 5.1 The vacuum spectrum.

Table 5.1 Summary of vacuum gauge performance

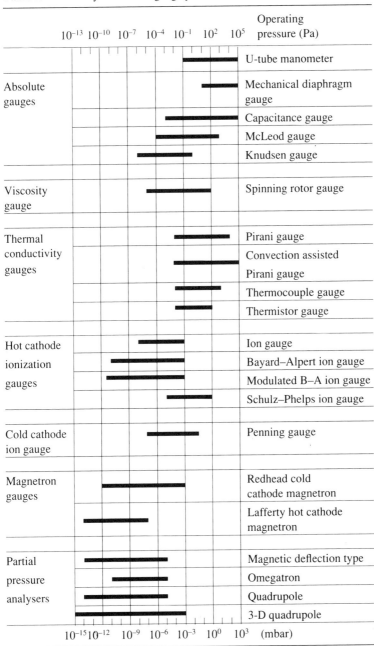

	10^{-13} 10^{-10} 10^{-7} 10^{-4} 10^{-1} 10^{2} 10^{5}	Operating pressure (Pa)
Absolute gauges		U-tube manometer
		Mechanical diaphragm gauge
		Capacitance gauge
		McLeod gauge
		Knudsen gauge
Viscosity gauge		Spinning rotor gauge
Thermal conductivity gauges		Pirani gauge
		Convection assisted Pirani gauge
		Thermocouple gauge
		Thermistor gauge
Hot cathode ionization gauges		Ion gauge
		Bayard–Alpert ion gauge
		Modulated B–A ion gauge
		Schulz–Phelps ion gauge
Cold cathode ion gauge		Penning gauge
Magnetron gauges		Redhead cold cathode magnetron
		Lafferty hot cathode magnetron
Partial pressure analysers		Magnetic deflection type
		Omegatron
		Quadrupole
		3-D quadrupole

10^{-15} 10^{-12} 10^{-9} 10^{-6} 10^{-3} 10^{0} 10^{3} (mbar) gauge will be required to span the range indicated. It should be noted that it is unusual for a given pressure gauge to have a measuring range in excess of nine decades, and only one gauge, the Redhead magnetron gauge, reaches this range with one unit. Accordingly, it is always a question of choosing a pressure gauge, or gauges, to match the range of pressures likely to be encountered in the experimental system. Usually, two different types of gauge will be required to give full operational coverage.

It is apparent that pressure measuring systems based on measuring a difference in mercury or oil levels, i.e. direct measurement of the hydrostatic head, must have very limited application, since detecting changes equivalent to a fraction of a millimetre is not easy. There are additional general problems associated with pressure measurements, which can be summarized as:

1. The gauge cannot always be placed at the point in the system where the pressure must be measured.
2. The calibration of the gauge will usually depend on the exact composition of the gas involved and this will not always be known.
3. The gauge itself may affect the local pressure by pumping action, or by adsorbing/desorbing gas.

The final problem associated with pressure measurement is the problem of setting a standard. Nearly all vacuum gauges have a calibration depending on the nature of the gas. It is essential, therefore, to calibrate such gauges against a standard. For pressures above 1 mbar, the mercury or oil manometer may be used, but for pressures below 1 mbar this is not satisfactory as has been pointed out, and resort must be had to the McLeod gauge in which the gas is compressed by a known ratio before its pressure is measured by a mercury manometer. The McLeod gauge is the only readily available absolute gauge in the pressure range down to 10^{-6} mbar, although its accuracy is no more than $\perp 10\%$ at 10^{-4} mbar, and less than this down to 10^{-6} mbar.

5.2 ABSOLUTE GAUGES

By an absolute gauge we mean a gauge that measures the total pressure but whose calibration is independent of the gas composition and depends only on geometric or mechanical factors. Within this description there are really only five gauges to choose from, namely:

1. the U-tube manometer;
2. the McLeod gauge;
3. the Wallace and Tiernan diaphragm gauge;
4. the capacitance gauge;
5. the Knudsen gauge.

The most important of these four gauges, and certainly the oldest gauge still in use, is the McLeod gauge, introduced by H. G. McLeod in 1874.

The U-tube manometer is, of course, very restricted in its sensitivity once the pressure falls to around 0.5 mbar. Many attempts have been made to improve its sensitivity and the Rayleigh gauge introduced in 1901 is an example of a simple, but effective, method of extending the range down to 10^{-3} mbar. Figure 5.2 shows the Rayleigh differential manometer. It consists of two glass bulbs, one of which connects via tube T to a good vacuum, $< 10^{-5}$ mbar, and the other to the pressure to be measured. Each bulb has a glass pointer P sealed into it and the bulbs are interconnected through a T-connector and thence to the upper end of a barometric column. Mercury or oil can be raised or lowered in the bulbs by means of a reservoir connected to the barometric column and the level brought up so as to be flush with the end of the reference pointer P_1. Gradual tilting of the framework upon which the bulbs are mounted enables the mercury level to be brought into contact with pointer P_2. The amount of tilt is measured with an optical lever system based on a mirror fastened to the two bulbs. The measured tilt then

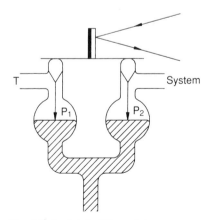

Fig. 5.2 Rayleigh differential manometer.

gives the difference in levels. Use of a low density vacuum oil, in place of mercury, is a useful way of improving the sensitivity.

The McLeod gauge is shown in Fig. 5.3(a). By raising the mercury with atmospheric pressure, the gas in the volume V is compressed. Its pressure is equal to the difference in height between the mercury column in the closed-off volume and the mercury column in the open tube which is connected to the vacuum system. There are two ways in which this gauge can be used. In the first, the mercury column in the open tube is always raised until it is level with the top of the closed capillary (Fig. 5.3(b)); the relationship between the pressure and the height h is

$$p_1 V = p_2 v = Ah^2 \tag{5.1}$$

Here p_2 is the pressure in the vacuum system, V is the volume which is closed off (bulb plus capillary), v is the volume remaining after compression and p_2 is the pressure in this volume, actually equal to p_1+h. The closed capillary has a cross sectional area A and h is the height of the gas column in the capillary. We can rewrite (5.1) as

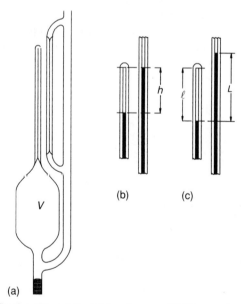

Fig. 5.3 (a) Section through the McLeod gauge. (b) The usual arrangement for pressure reading. (c) An alternative arrangement for pressure reading.

$$p_1 = \frac{A}{(V - Ah)} h^2 \qquad (5.2)$$

which shows that p_2 varies with h^2 through a geometrical constant A/V if we neglect the term Ah in the denominator; the error from this omission can be made $< 1\%$.

A more useful approach, which gives a larger pressure range, is the following (Fig. 5.3(c)). The mercury is raised in the closed capillary to a predetermined fixed point, a distance ℓ from the top. The difference in height between the columns is then L. Once again we can write

$$p_1 V = A\ell L$$

which gives us

$$p_1 = \frac{A\ell}{V} L \qquad (5.3)$$

Here $A\ell/V$ is a geometrical constant of the instrument and p_1 is directly proportional to L.

Although the operation of this gauge is conceptually simple, there are various precautions and restrictions which must be observed. Thus, the gauge cannot be used as described for the measurement of vapours, since these would condense during compression and in any case depart from the ideal gas law. It is usually necessary to provide a liquid nitrogen cold trap to prevent mercury from condensing in the vacuum system. Pressure can only be measured intermittently rather than continuously. The gauge is fragile; it is quite possible to cause the capillary to be blown off the bulb by allowing the mercury level to rise too rapidly in the bulb. Although a low pressure limit of 10^{-6} mbar is set by the compression ratio, the actual limit is nearer to 10^{-4} mbar owing to geometrical uncertainties, reading errors and dirty glassware. A further error in low pressure measurements is set by mercury vapour streaming from the gauge into the adjacent cold trap. In order to reduce this error to less than 5%, the tube connecting the gauge to the system should not exceed 8 mm internal diameter.

A convenient and compact version of the McLeod gauge is the Vacustat, a swivel-type McLeod gauge based on a design introduced by E. W. Flosdorf in 1938. It is very convenient for measuring gas pressures in the range from 10^{-2}–10 mbar, or 1–10^{-3} mbar, according to the head chosen. It requires only about 140 g of mercury and avoids the use of a movable reservoir and auxiliary vacuum pump. Reference

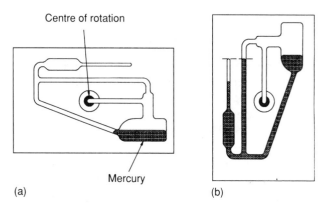

Fig. 5.4 The Vacustat (Edwards High Vacuum Ltd). The vacuum system is connected to the gauge at the centre of rotation.

to Fig. 5.4 will clarify the mode of operation. The gauge head is normally positioned with the measuring capillary and scale horizontal (Fig. 5.4(a)). To take a measurement, the head is rotated through 90° around its rotary seal to bring the tube and scale upright (Fig. 5.4(b)). The rotation causes the mercury to move within the glassware and isolate a fixed volume of the gas whose pressure is to be measured. The trapped gas is then compressed into the closed capillary tube and when the mercury in the open capillary is level with the zero mark on the measuring scale, the system pressure is given by the height of the mercury column in the closed capillary measured against the scale which is calibrated directly in mbar.

An alternative approach to absolute pressure measurement is that depending on the force operating on a sealed diaphragm. In the Wallace and Tiernan version of this approach, pressure measurement is accomplished by admitting the unknown pressure to the hermetically sealed instrument case, where it exerts a pressure on a flat, evacuated capsule that has been permanently sealed. Movement, expansion or contraction of the capsule is transmitted by a lever system to the pointer that registers the pressure on the dial as a pressure above zero. The elements of the gauge are shown in Fig. 5.5.

An elegant variation of the diaphragm gauge described above is the capacitive manometer in which diaphragm movement is determined electrically rather than mechanically. This approach yields high accuracy and extreme stability. Figure 5.6 shows a section through such a gauge head. Basically it comprises a sealed metal housing, A, which is

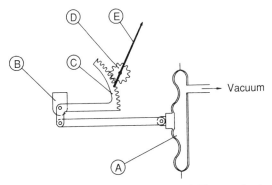

Fig. 5.5 The mechanical basis of the Wallace and Tiernan absolute pressure indicator; A is the pressure sensitive element; B is the pivot point; C is the geared sector; D is the pinion and E is the pointer.

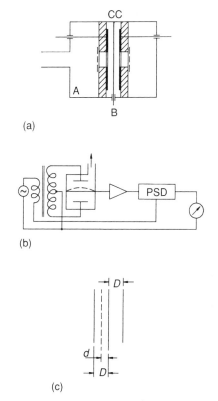

Fig. 5.6 (a) Cross section through a capacitance manometer based on the Barocel design. (b) Excitation circuitry. (c) Electrode arrangement. (The Barocel is marketed by Edwards High Vacuum).

divided into two identical sections by a thin, radially prestressed, metal diaphragm, B. This diaphragm is positioned symmetrically between fixed capacitor plates, C, formed on ceramic discs. The net result is a capacitive potential divider in which minute motion of the central diaphragm, following a pressure change, varies the relative capacitance of the diaphragm and the fixed capacitor plates. The plates form part of a bridge circuit (Fig. 5.6(b)), excited by a 10 kHz signal, so that any displacement of the diaphragm unbalances the bridge and produces a 10 kHz signal of amplitude proportional to the pressure. Again, the output signal is a measure of total pressure and is unaffected by variations in gas composition.

Since the construction is absolutely symmetrical both mechanically and electrically, temperature variations have little effect, giving very high zero stability. Also, because the sensor is a pure capacitive device, high (16 V) excitation voltages may be used, thus minimizing signal amplification requirements. Pressure changes as small as 1×10^{-7} of the full range of the sensor can be detected, with a linearity approaching 0.01%.

The balanced·gauge structure allows for either differential measurements, with pressure changes applied to both sides of the diaphragm, or straightforward pressure measurement against a self-contained vacuum reference, obtained by sealing off one side of the diaphragm at high vacuum. Considering electrical elements in Fig. 5.6(c) we note that for diaphragm separations $(D + d)$ and $(D - d)$ we can write that

$$\frac{C_{out}}{C_{in}} = \frac{d}{D} \tag{5.4}$$

and if the displacement d is proportional to pressure then

$$\frac{C_{out}}{C_{in}} = const.p \tag{5.5}$$

yielding a linear capacitance variation with pressure. In more rigorous treatment of the diaphragm displacement the relationship is not perfectly linear.

The capacitance manometer structure may be formed from materials which are compatible with corrosive gases such as F_2, UF_6, Cl_2 and will normally provide four decades of pressure measurement within one pressure cell. The range indicated in Table 5.1 can only be achieved with two gauges.

An interesting, but rather less frequently used absolute gauge, is the Knudsen gauge of 1910, which depends on momentum transfer in gases at low pressures. Knudsen showed that a mechanical force is exerted between two surfaces maintained at different temperatures in a gas at low pressure. Reference to section 1.12 will show that molecules striking a hot surface rebound with a higher average kinetic energy than those which strike a cooler surface. Figure 5.7 shows a diagrammatic sketch of the Knudsen type gauge. The two parallel strips, A and B, are placed at a distance apart which is less than the mean free path of the molecules. The strip A is at the temperature T_0 of the residual gas, while B is maintained at a higher temperature T_1. On the side away from B, A will be bombarded by molecules having root mean square velocity c_0 corresponding to the temperature T_0, as given by the equation

$$c_0 = \left(\frac{3RT_0}{M} \right)^{1/2}$$

Molecules leaving A will have the same velocity. On the side towards B, however, A will be bombarded by molecules coming from B and

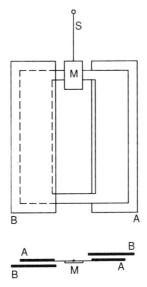

Fig. 5.7 Diagrammatic sketch of the Knudsen-type pressure gauge.

having a higher velocity c_1 corresponding to T_1. As a result A will receive momentum at a greater rate on the side towards B than on the opposite side, and will therefore experience a torque.

Put in more precise terms the net force F which is exerted is given by the rate at which momentum is transferred per unit area from B to A, thus we can write

$$F = \frac{1}{6}n_0\left(mc_1 - mc_0\right) \qquad (5.6)$$

In terms of the pressure around the strips A and B we have

$$p = 4F\left(\frac{T_0}{T_1 - T_0}\right) \qquad (5.7)$$

If the temperatures T_0 and T_1 are maintained constant, then the force F is directly proportional to the pressure and is independent of the molecular weight of the gas. The force F is conveniently measured by recording the twist occurring on the torsion fibre S. This is usually done optically. Although this device is an absolute manometer in the sense that it does not need calibration against another gauge, it is rather fragile, and despite being an elegant example of the application of kinetic theory at low pressures and a demonstration of momentum transport by gas molecules, it does not now see much use except in the calibration of other gauges. The useful operating range of this gauge lies in the range 10^{-2}–10^{-8} mbar although it operates best over a more limited range 10^{-4}–10^{-7} mbar. At the low pressure end of its range it is markedly superior to the Pirani gauge which has displaced it.

5.3 SPINNING ROTOR GAS FRICTION GAUGE

The spinning rotor gas friction gauge is based on an idea first demonstrated by J. W. Beams in 1962. A small steel ball bearing, approximately 4.5 mm in diameter, is suspended magnetically in a vacuum and spun at 410 revolutions per second by a rotating magnetic field produced by four drive coils. The drive and stabilization coils are set up around the stainless steel vacuum tube in which the ball is sited. For glass vacuum systems a glass tube may be used. The physical arrangement of coils, etc., is shown in Fig. 5.8.

After switching off the drive signal the rotational speed of the ball decreases as a function of the pressure due to gas friction (viscosity). In

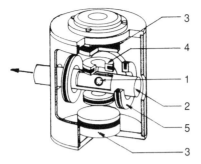

Fig. 5.8 Cut away view of the Leybold spinning rotor gas friction gauge, with permission; (1) is the Spinning ball; (2) is the vacuum tube; (3) are two permanent magnets; (4) are two vertical stabilization coils and (5) is one of the four drive coils. This view does not show the four lateral stabilization coils or the four pick-up coils required.

practice the pressure is determined by computation based on the measured deceleration (detected by four pick-up coils), and the parameters for the gas and the spinning ball entered by the user. The operator may select between fixed measurement times or optimally selected measuring times. In any event the pressure is displayed digitally.

The pressure is related to the ball and gas parameters by

$$p = \frac{1}{\sigma_m} \frac{a\rho}{5} \frac{\pi\bar{c}}{2} \frac{\tau_{n+1} - \tau_n}{\tau_{n+1}\tau_n} \tag{5.8}$$

where p is the gas pressure, σ_m is the coefficient of tangential momentum transfer (very close to unity for all gases), a is ball radius, ρ is ball density, $\bar{c}$ is mean molecular velocity and τ_n, τ_{n+1} denote two successive time intervals required for a certain number of rotor revolutions.

There are several features of the spinning rotor gauge which make it attractive:

1. It is very accurate, better than 3%, in the pressure range 10^{-7}–1 mbar (superior to thermal conductivity and ionization gauges).
2. A given, calibrated, ball may be transferred to another vacuum system without destroying the calibration. It is, in effect, a transferable standard.
3. There are no gas evolution or adsorption processes which can affect the pressure measurement.

It is not, of course, an absolute gauge since for accurate measurement of pressure it is necessary to know the gas composition in order to calculate $\bar{c}$.

5.4 PIRANI GAUGE

For a given gas at sufficiently low pressures, the thermal conductivity decreases with pressure, as described in section 1.13. This concept is used in a group of gauges which have a means of sensing the heat loss from a hot source through the surrounding gas. Perhaps the simplest form of these gauges is the Pirani gauge (1906), in which a thin wire of diameter d is placed along the axis of a tube of much larger diameter D, the open end of which is connected to the system in which the gas pressure p is to be measured. The axial wire is heated to around 200° C by the passage of electric current. As the gas pressure around the wire changes, the heat loss from the wire changes also, so that the temperature of the wire rises as the pressure falls and falls as the pressure rises. This variation of the temperature of the wire with pressure is reflected by a variation in the resistance of the wire with pressure. Consequently, the pressure may be measured by incorporating the gauge into a Wheatstone bridge.

The theoretical basis of the Pirani gauge can be assembled reasonably easily. Suppose the central wire is heated by the passage of electric current to a temperature T (K), whilst the inner surface of the outer tube is maintained at the ambient temperature T_A (K), $T > T_A$. Now, the diameter of the wire is very much less than the diameter of the tube, typically 0.05–0.1 mm compared with 10 mm, so that gas molecules in the gauge tube will make many more collisions with the tube than with the wire. Consequently, we can assume that the gas molecules are at a temperature T_A and only reach the higher temperature when they strike the wire. We assume also that after collision with the wire, the molecules attain an energy corresponding to the temperature T.

The rate of transfer of thermal energy E from the heated wire to the gas is given by

$$E = vk(T - T_A)$$

where v is the molecular collision frequency corresponding to T_A and k is a constant chosen to suit a given gas and wire. If the distance travelled by the average impinging molecule is $< \lambda$, the mean free path, we can substitute for v, using (1.49), to find that

$$E = \frac{kpA(T - T_A)}{(2\pi m k_B T_A)^{1/2}} \tag{5.9}$$

where A is the surface area of the wire. This equation implies that all colliding molecules have the wall temperature T_A. As the pressure around the wire increases, the mean free path λ will decrease until it becomes less than the wire diameter d. Under these conditions, many molecules will return to the wire immediately after leaving it, thus invalidating (5.9); the heat transfer per unit time becomes independent of pressure. Equation (5.9) is approximately correct provided that $\lambda > d$, so that for nitrogen gas and a typical wire diameter of 0.01 mm, the limiting pressure is around 0.7 mbar.

We can see from (5.9) that the heat transfer from the wire at a given pressure p will vary inversely with the square root of the molecular mass m. It will also depend on k, which varies with the nature of the gas and the surface condition of the wire.

While heat transfer through the gas is the means by which pressure is measured in the Pirani gauge, there are three other sources of heat loss from the wire, namely, radiation, convection and conduction through the metal leads to the ends of the wire. However, radiation and conduction through the leads are independent of pressure. Heat loss by radiation becomes dominant once the pressure falls below about 10^{-5} mbar, whilst conduction losses will be less the longer the wire filament used. Once the filament is lengthened past a certain point, it becomes impossible to stabilize it mechanically and the consequent vibrations introduce additional gas/wire energy changes. The net effect of these two factors is that the lower limit of Pirani gauge operation is about 10^{-4} mbar.

The upper pressure limit of a thermal conductivity gauge is determined by the saturation pressure of the thermal conductivity. This occurs at a Knudsen number of about 10^{-2}, Fig. 5.9. In consequence, most thermal conductivity gauges have an upper pressure limit of about 10^{2} mbar; 100 mbar. however, the design may be altered to encourage and make use of convection losses, which operate at higher pressures, and this has the effect of extending the available pressure range to 10^{3} mbar. A gauge embodying this idea was introduced by Granville–Phillips in 1977 and is marketed as the Convectron; alternative versions are available from other manufacturers although the operating principles are the same.

The practical implementation of the Pirani gauge generally takes the form of a glass tube with a diameter of 10 mm, having an O-ring

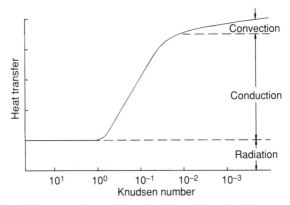

Fig. 5.9 Heat transfers regimes for a thermal conductivity gauge.

connection or greased cone joint at its open end, that is, the end to be connected to the vacuum system. The Pirani filament is of tungsten, nickel or platinum wire, 0.05–0.1 mm in diameter, wound into a helix of axial length 8–10 cm and outside diameter 0.5–2 mm. The pitch of the helix is set at about 1 mm and the total length of wire involved is some 20–30 cm, Fig. 5.10(a). Generally, a physically identical filament is mounted in close proximity, but in a tube sealed to high vacuum. This filament acts as a compensating head since it is subject to the same ambient temperature fluctuations. It is connected in the opposing arm of the Wheatstone bridge measuring circuit (Fig. 5.10(b)). In practice the usual arrangement is for the bridge to be supplied from a constant voltage source and the bridge is balanced by adjusting R_V in Fig. 5.10(b) until the meter reading is zero when the ambient pressure is less than 10^{-4} mbar. Consequently, as the pressure rises the micro-ammeter M records an out-of-balance signal which is related to pressure through a calibration peculiar to the particular gauge. This calibration is not linear with pressure.

Commercially available, mass-produced, Pirani gauge heads and controllers come with the meter M precalibrated for dry air. This calibration may be temporarily lost if the gauge is exposed to 'dirty' conditions, i.e. water vapour or oil. It is good vacuum practice to prevent this type of exposure by suitable trapping, as described in Chapter 6. Among the gauges which depend on thermal conductivity, the Pirani gauge has a response to pressure changes which is ten times

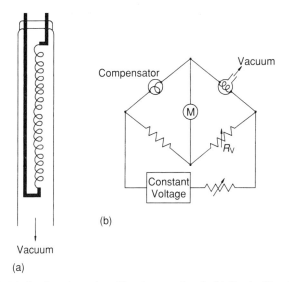

Fig. 5.10 (a) Section through a Pirani gauge head. (b) Basic Pirani gauge control circuit.

faster than, for example, the thermocouple gauge. Any given Pirani head is unlikely to cover a pressure range extending over more than five orders of magnitude, so that it is generally necessary to use more than one gauge head if a larger pressure range is required.

5.5 THERMOCOUPLE GAUGE

The thermocouple gauge mentioned above operates exactly like the Pirani gauge, the difference being the means selected to measure the temperature change of the wire as the pressure alters. In this case, a thermocouple is attached to the centre of the heated wire, Fig. 5.11, which is, in turn, supplied from a simple, constant current source. A moving coil meter records the thermocouple e.m.f. directly, although it is a simple matter to amplify this output voltage if required. The pressure range embraced by the thermocouple gauge extends from 10–10^{-4} mbar, although any one instrument is unlikely to cover more than four orders of magnitude. Once again the calibration is not linear with pressure change and is, of course, different for different gases. The thermocouple gauge was first introduced by W. Voege in 1906.

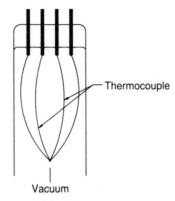

Fig. 5.11 A thermocouple gauge.

5.6 THERMISTOR GAUGE

An alternative arrangement to that employed in the Pirani and thermo-couple gauges is to use a thermistor run so as to be self heating: this approach was first demonstrated by W. Meyer in 1938. The thermistor usually chosen for this purpose is a miniature glass-covered bead type, mounted on support leads sealed into the vacuum envelope (Fig. 5.12).

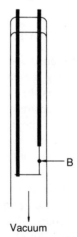

Fig. 5.12 A thermistor gauge head; B is a glass-coated bead thermistor.

The thermistor is heated to around 100° C by passage of a current. The resistance of a thermistor is an exponential function of temperature change so that the current which must be supplied to the thermistor to maintain constant resistance, and hence constant temperature, is itself a sensitive function of temperature and hence also pressure. The thermistor is incorporated into a Wheatstone bridge which may either be operated to be permanently balanced, by altering and recording the bridge supply voltage, or may be run with a constant bridge supply voltage with the out-of-balance voltage being recorded.

As a family of pressure gauges, thermal conductivity gauges have some useful advantages:

1. The time lag between pressure change and gauge response is negligible.
2. Units are compact and may be operated remotely.
3. They indicate total pressure, i.e. permanent and condensable vapours. Contrast the McLeod gauge which records only permanent gases.

The disadvantages which attend thermal conductivity gauges are:

1. In general they have a lower pressure limit of 10^{-4} mbar.
2. The gauge calibration varies with the nature of the gas, varying with $\sqrt{M}$.
3. The gauge is sensitive to contamination which will alter the calibration. This effect can be minimized by using the glass-covered thermistor bead type.

Despite the disadvantages listed above, the thermal conductivity gauge, generally in the form of a Pirani gauge, is the usual choice for monitoring the backing pressure in a vacuum system.

5.7 IONIZATION GAUGES

One of the largest, most important and most widely used class of pressure gauge is the hot cathode ionization gauge. Its importance stems from the fact that, by appropriate choice of gauge, measurements may be conducted over the pressure range $1-10^{-11}$ mbar, a greater range than is available with any other gauge type.

Now the kinetic energy acquired by an electron in passing through a potential difference of V volts is just Ve, where e is the charge of the

electron. If this energy exceeds a certain critical value, corresponding
to the ionization potential of the gas V_i, there is a definite probability,
which varies with V and the nature of the gas, that a collision between
the electron and a gas molecule will detach an electron from the gas
molecule to form a positively charged ion. The range of values for V_i
runs from 24.58 volts for helium with the highest ionization potential,
down to 3.88 volts for caesium which has the lowest. For the common
constituent gases of a vacuum system, e.g. O_2, N_2, the value is around
15 volts.

In a thermionic diode structure with a fixed anode voltage greater
than V_i, the number of positive ions produced by electrons travelling to
the anode should be directly proportional to the pressure, provided that
the pressure is sufficiently low so that any one electron does not make
more than one ionizing collision during its transit. If the pressure of
gas is kept constant, then the number of positive ions produced will
depend on the number of electrons emitted by the cathode, provided
that space charge effects are negligible and do not alter the potential
distribution of the diode.

On the basis of the physical ideas sketched out above, we can see
that this arrangement can be made to constitute a pressure measuring
system if we keep the electron emission constant and add a third
electrode, negatively charged with respect to the cathode, with which to
measure the ion current. In reality we cannot expect to collect every
positive ion which is formed, only a constant fraction of them. This
three-electrode device is called an ionization gauge and it exists in a
variety of forms, each designed to meet a particular objective.

If we denote the positive-ion current to the ion collector by i_p, and
the electron-emission current to the anode by i_e, then if the pressure in
the ionization gauge is p, we can write that

$$i_p = S' i_e p \qquad (5.10)$$

where S' is a proportionality constant known as the sensitivity of the
gauge. Provided that we measure i_p and i_e in the same units, then S' has
the units of reciprocal pressure.

It is instructive to substitute some values in (5.10) to gain some idea
of the efficiency of the ionization process. Thus, if we have a gauge
with a sensitivity of 20 mbar^{-1}, then at 10^{-6} mbar we find that only one
ion is produced for 50 000 electrons emitted from the cathode. The
probability of ionizing a gas by electron impact is actually a function of
the electron energy. Figure 5.13 shows how this probability varies with
energy and that it varies from one gas to another. It is actually plotted

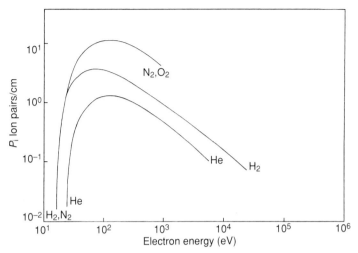

Fig. 5.13 Ionization probability versus electron energy for some common vacuum system gases.

as P_i, the number of ion pairs (a singly charged positive ion and one electron) produced per centimetre of path length by an electron travelling through a gas at 1 mbar pressure and $0°\,C$, against the electron energy in electron volts.

The definition of P_i above allows us to write that

$$N_i = P_i L p \qquad (5.11)$$

where N_i is the number of ions pairs produced, L is the path length travelled by an electron, and p is the gas pressure. It is tempting to call P_i the probability of ionization, but it is really a differential ionization coefficient, so that when a constant value of P_i, corresponding to the initial kinetic energy of an electron, is substituted in (5.1), this equation can only be used when $N_i \leq 1$. If $N_i = 1$, then L becomes the length of the mean free path travelled by an electron before making its first ionizing impact. After its first ionizing impact the electron will have less kinetic energy and a new value of P_i will be required in order to calculate the average distance before the next ionizing collision. If we assume that the number of ions produced per electron is just the ratio of the ion current to the electron current, then we can substitute in (5.11) to get

$$\frac{i_p}{i_e} = P_i L p \qquad (5.12)$$

This assumption neglects the fact that there are more ions formed in the gauge than reach the ion collector and, of course, P_i is not a constant over the electron path length. Nevertheless, if we use a value for P_i corresponding to the electron energy at the anode potential, we can get some feel for the effective path length travelled by the electrons in moving from cathode to anode, since this is now what L will represent. Substitution of (5.12) into (5.10) gives

$$L = \frac{S'}{P_i} \qquad (5.13)$$

If one substitutes typical values for S' and P_i corresponding, say, to nitrogen, then at 10^{-6} mbar the effective path length of electrons in an ionization gauge is around 20 mm, yet the electron mean free path computed from (5.11) using $N_i = 1$ is about 900 m. This implies that at 10^{-6} mbar, most of the electrons make no ionizing collisions during their flight from cathode to anode and only a few make even one. It is worth noting at this juncture that any means of artificially increasing the electron path length will have an interesting and useful effect on gauge performance. Gauges embodying such means will be described in sections 5.12, 5.13 and 5.14.

5.8 THE HOT CATHODE IONIZATION GAUGE

The first ionization gauge built upon the principles outlined above was introduced in 1916 by O. E. Buckley. In its original form it is essentially a triode valve. Electrons from a central heated filament are accelerated towards an electrode maintained at a positive potential in excess of that corresponding to the peak of the ionization probability curves of Fig. 5.13, i.e. somewhat in excess of 100 volts. The anode takes the form of a cylinder enclosing the electron-emitting filament and the ion collector grid, Fig. 5.14. The ion collector grid is held at a potential of around 20 V, negative with respect to the filament. The whole electrode structure is usually sealed into a glass envelope connected to the system in which the pressure is to be measured. With the electrode potentials specified, the pressure in the gauge is a linear function of the ion current provided that the emission current is maintained constant. There is an additional problem with this type of gauge in that the hot filament and the electron emission current cause gas evolution from the gauge structure. Fortunately, this is easily

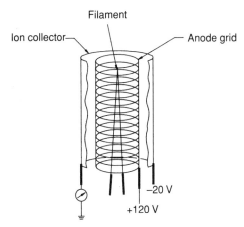

Fig. 5.14 The electrode arrangement of conventional triode-type hot cathode ionization gauge.

remedied by increasing the electron emission above its normal value and changing the anode potential to around 1 kV, when the gauge structure may be raised to red heat by electron bombardment; the gauge structure then outgasses.

In operation, the hot cathode ionization gauge described above cannot record pressures below 10^{-8} mbar. The reasons for this were first suggested in 1947 by W. B. Nottingham, who proposed that the residual current was not due to the collection of positive ions and was thus completely independent of pressure. As a consequence of this pressure limitation the conventional hot cathode ionization gauge has been essentially replaced by an improved, or inverted, gauge due to R. T. Bayard and D. Albert and introduced in 1950.

5.9 THE BAYARD–ALPERT GAUGE

The residual current found in a conventional ionization gauge is caused by photoelectrons ejected from the ion collector by soft X-rays produced by 100–200 V electrons striking the cylindrical anode grid. In terms of the external measuring system, photoelectrons emitted from the ion collector cannot be distinguished from positive ions incident on the collector and there must exist a lower limit to the value of i_p with decrease in pressure, which corresponds to pure photoelectric emission.

For conventional gauges, at normal operating voltages, this X-ray photocurrent is about 2×10^{-7} times the anode current. Thus, for a gauge with a sensitivity of $S' = 20$ mbar^{-1}, the ion current will be just equal to the photoelectric current at a pressure of 10^{-8} mbar.

The Bayard–Alpert design overcomes this problem by inverting or exchanging the positions of the filament and ion collector. The filament is now placed outside the cylindrical anode grid, and the ion collector, which is now a fine wire rather than a large area cylinder, is suspended at the centre of the anode grid (Fig. 5.15). The usual potentials are applied to the gauge electrodes so that electrons from the filament are accelerated into the grid cylinder where they make ionizing collisions. A large fraction of these ions are collected by the centre wire. Now, however, the solid angle presented by the ion collector to the X-rays emitted by the anode grid has been reduced by a factor of several hundred with the consequence that, although the same number of X-ray photoelectrons are emitted, many fewer are collected. The outcome is that the pressure limit on the gauge goes down to around 2×10^{-11} mbar and, for the first time, we have a gauge capable of operating in the ultrahigh vacuum range.

Another feature of the Bayard–Alpert gauge is the logarithmic potential distribution within the anode grid. This means that the potential within the grid is nearly uniform, except in the immediate vicinity of the ion collector wire. In consequence, electrons travelling within the anode grid have an efficient ionizing energy for most of the grid volume. In the conventional gauge, the potential falls more or less linearly between the anode grid and cylindrical ion collector and many electrons are decelerated to energies below the threshold for efficient

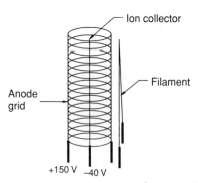

Fig. 5.15 The electrode arrangement of the Bayard–Alpert inverted hot cathode ionization gauge.

ionization. A further advantage of the Bayard–Alpert configuration is that it is possible to insert an additional, or stand-by, filament within the gauge envelope.

The Bayard–Alpert configuration of the hot cathode ionization gauge has effectively displaced the normal triode configuration from modern-day use. A wide range of gauges are now manufactured to the Bayard–Alpert design, either mounted in glass envelopes or nude on a stainless steel flange. Most manufacturers provide facilities for stabilizing the electron emission current at the values 10 mA, 1 mA and 100 μA. The choice of electron emission setting is important since all ionization gauges exhibit a pumping action which has chemical, as well as electrical, components. The former is due to gas reacting at the hot filament, whilst the latter is due to positive ions which are sorbed or buried on reaching the ion collector or the outer envelope of the gauge. The lower the emission current the lower the pumping action.

Since gauge heads pump gas, they may also under some circumstances evolve gas. It is important that they be attached to the vacuum system with large diameter tubulation if accurate pressure readings of the system pressure are required. With the nude head design, this problem does not arise since the gauge may be immersed directly in the vacuum chamber.

The chemical reactivity of the hot filament may be reduced by lowering its temperature, of course, but if electron emission is to be maintained at its previous level, then the filament work function must also be reduced. The usual solution to this problem is to use a chemically inert material such as rhenium, coated with lanthanum hexaboride (LaB_6). The effect of this coating is to enable the filament temperature to be reduced by about 1000° C whilst still maintaining an adequate emission current. (The filament work function is reduced to around 2.0 eV.)

5.10 MODULATED BAYARD–ALPERT GAUGE

In 1960, P. A. Redhead introduced a simple modification to the standard Bayard–Alpert gauge which enables the X-ray photoelectron current to be measured. To the standard Bayard–Alpert gauge structure is added a second ion collector wire within the electron collector grid, Fig. 5.16. The second collector, or modulator, can be switched to anode or earth potential at the gauge control unit. With the modulator at ground potential it will collect some of the positive ions, but not affect

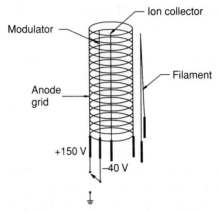

Fig. 5.16 The Bayard–Alpert gauge with modulator electrode.

the X-rays falling on the ion collector, so that the new total collector current is now given by

$$i'_c = i_x + i'_p \tag{5.14}$$

where i_x is the X-ray current. With the modulator at the anode potential, all positive ions reach the ion collector and the measured collector current is

$$i_c = i_x + i_p \tag{5.15}$$

For a given gauge, the ratio between i'_p and i_p is fixed and is known as the modulation factor α. It is determined at pressures above 10^{-8} mbar where the X-ray photocurrent is negligible compared with the ion current. Under these conditions we have

$$\alpha = \frac{i'_p}{i_p} = \frac{i'_c}{i_c}$$

A value of around 0.6 is typical. We can now write an expression for the X-ray current using (5.14) and (5.15)

$$i_x = \frac{i'_c - \alpha i_c}{1 - \alpha} \tag{5.16}$$

and thus arrive at an expression for the true ion current

$$i_p = \frac{i_c - i'_c}{1 - \alpha} \tag{5.17}$$

By means of the modulator electrode, the range of a conventional Bayard–Alpert gauge is extended down to 5×10^{-12} mbar.

5.11 THE HIGH PRESSURE ION GAUGE

It will be seen from table 5.1 that the high pressure limit of a conventional ion gauge or a Bayard–Alpert gauge is 10^{-3} mbar. Above this pressure the relationship between pressure and ion current ceases to be linear, indeed the ion current tends to become constant.
There are three reasons for this:

1. At high pressures the mean free path of electrons in the gas becomes comparable with the cathode–anode grid distance. Electrons now tend to lose too much energy by collisions so that they are no longer able to ionize gas molecules.
2. In the ionizing collision an ion pair is produced in which a low energy electron is formed that contributes to the measured emission current, but does not produce significantly more ionization. Since the total emission current is controlled electronically and held at a constant value, the effect of these low energy electrons is to reduce the number of electrons emitted by the hot filament; the gauge sensitivity then appears to decrease.
3. A positive ion space charge begins to form in the filament region, accompanied by a glow discharge when the pressure reaches a few mbar. The plasma accompanying the glow discharge scatters ions away from the ion collector.

These effects are minimized in the Schulz–Phelps gauge (1957) to such an extent that the high pressure limit is raised to 1 mbar; the low pressure limit also rises, to 10^{-5} mbar. The electrode arrangements are shown in Fig. 5.17 and comprise a straight iridium filament set equidistant from, and parallel to, two identical parallel molybdenum plates. The spacing between each plate and the filament is 1.5 mm while the plates themselves measure 9×12 mm. One of these plates is held at +60V and acts as the electron collector; the other is the ion collector operated at −60V.

Because the distance between filament and anode is now rather small and the accelerating potential reduced, electrons on average make only one collision on moving from cathode to anode, thus reducing the contribution from (1) but at the sacrifice of much gauge sensitivity. By the same token (2) is also reduced so that electrons from ion pairs

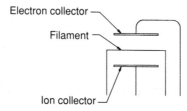

Electron collector
Filament
Ion collector

Fig. 5.17 The Schulz and Phelps hot cathode ionization gauge.

constitute less than 10% of the total electron current. Factor (3) is controlled by virtue of the fact that the ion collector area is large compared with the filament and operates at a high negative potential. Unfortunately, the large surface area collector ensures that the X-ray limit is raised to $\sim 10^{-6}$ mbar.

5.12 THE PENNING GAUGE, COLD CATHODE IONIZATION GAUGE

It was pointed out in section 5.7 that the ionization current produced in a gas by the application of a high voltage depends on the gas pressure, and further, it is not possible to sustain sufficient ionization of the gas at low pressures unless the electron paths are greatly increased by some means. The usual means employed to increase the path length is to confine the discharge with a magnetic field which has the effect of causing the electrons to spiral and oscillate on greatly increased paths.

The original cold cathode ionization gauge using this form of confinement was introduced by F. M. Penning in 1937. A pair of identical cathode plates are set equidistant either side of an anode ring within a glass envelope, and an axial magnetic field strength of about 32×10^3 Am^{-1} is maintained to confine the discharge. A permanent magnet supplies the magnetic field. With the anode potential set at 2 kV, a significant ionization current may be recorded down to 10^{-5} mbar, since the electrons initially produced in the discharge and those resulting from positive ion impact on the cathode now execute very long paths, several hundred times longer than the anode–cathode clearance.

The Penning gauge is now available with both glass and metal envelopes and provides a simple, rugged gauge for pressure measurement in the range 5×10^{-8} to 10^{-2} mbar. Nevertheless, the readings

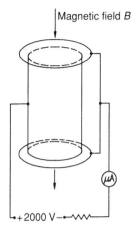

Fig. 5.18 Perspective view of electrode structure of cold cathode ionization gauge of the Penning–Nienhuis type.

obtained from it must be used with some caution since the gauge exhibits a much greater pumping action than a conventional ionization gauge, and must be recalibrated from time to time for individual gases. A further difficulty exhibited by this gauge is the problem of starting. Some means must be available to initiate the discharge; often this will be provided by an incident energetic particle (cosmic ray), but in some gauges a sharp point is attached to the cathode to produce electrons by field emission. In any event starting at low pressures is problematical. Generally speaking, the modern version of the Penning gauge employs a somewhat different electrode geometry with a cylindrical anode instead of a ring, and a higher magnetic field strength of $80 \times 10^3 \, \text{Am}^{-1}$. This reduces calibration changes and other irreproducible behaviour, and extends the pressure range down to 5×10^{-8} mbar. It was first described by Penning and Nienhuis in 1949, Fig. 5.18.

5.13 MAGNETRON IONIZATION GAUGES

The cold cathode ionization gauge has been further developed by P. A. Redhead to produce a gauge which spans the pressure range from 10^{-3}–10^{-12} mbar. This gauge is known as the 'inverted magnetron' or Redhead gauge and is available commercially. The name arises from its similarity in structure to the magnetron valve, with a central axial electrode surrounded by a separate cylindrical electrode. Here, however, the central electrode is the anode and the outer one the cathode, thus inverting the conventional magnetron structure.

In a cold cathode ionization gauge, the lower pressure limit may, in principle, be extended by increasing the applied potential so that it is considerably greater than 2 kV. However, as the potential is increased, although operation at low pressure is enhanced, the ion current reaches a limiting value set by field emission of electrons from the cathode. This sets a lower pressure limit for the operation of the Penning gauge, analogous to the X-ray limit for the Bayard–Alpert gauge.

In the inverted magnetron gauge, a cold cathode discharge is established in crossed electric and magnetic fields. A view through such a gauge is shown in Fig. 5.19. The gauge comprises a central anode wire, held at a potential of 5–10 kV, surrounded by a closed cylindrical ion collector, in turn surrounded by a closed auxiliary cathode. The auxiliary cathode has three functions: it acts as a shield for the ion collector and, by means of two short tubular shields which project 2 mm into the ion collector, provides the field emission which initiates the discharge. Lastly, these same two tubular shields protect the ion collector end-plates from the high electric field. A permanent magnet sleeve provides the axial magnetic field of 160×10^3 Am^{-1}. In operation the ion current is measured between the ion collector and the anode with the auxiliary cathode grounded. The field emission current is now largely confined to the auxiliary cathodes and is therefore not recorded, whilst the magnetic field ensures efficient electron trapping in the discharge region.

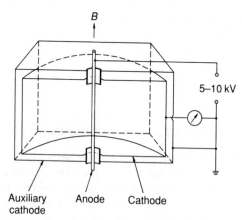

Fig. 5.19 Cut away view of the Redhead inverted magnetron gauge.

The ion current is related to the pressure by the equation

$$i_p = kp^n \qquad (5.18)$$

where k is a constant, depending on the nature of the gas, and n is a constant depending on the gauge: it typically has a value between 1.1 and 1.4 and depends on the magnetic field strength chosen. At high magnetic fields it approaches unity. The sensitivity or conversion constant k, of this gauge is about 1 A mbar^{-1}. Note in these gauges that the sensitivity is no longer in units of reciprocal pressure.

5.14 HOT CATHODE MAGNETRON GAUGE

There is another member of the magnetron ionization gauge family which should be mentioned, and that is the hot cathode version due to J. M. Lafferty and first described in 1961. The approach is to increase the electron path length as it travels from cathode to anode, using a magnetic field. The gauge is shown in cut-away form in Fig. 5.20. Its structure is reminiscent of the conventional ion gauge inasmuch as the filament lies at the centre of a cylindrical anode, but there the similarities end since the ion collector is a circular disc which closes off the top of the anode, whilst the bottom of the anode is closed by another negatively-biassed shield.

In the Lafferty gauge, the low pressure limit of the conventional hot cathode ion gauge is extended by the simple expedient of increasing

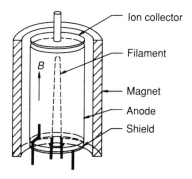

Fig. 5.20 Cut away view of the Lafferty hot cathode magnetron gauge. The gauge vacuum envelope is omitted for clarity; it would fit just inside the cylindrical permanent magnet.

the ratio of the ion current to the X-ray photocurrent, at a given electron emission current and gas pressure. In effect, the gauge sensitivity is increased by causing the electrons to travel in long spiral paths about the cathode. If the magnetic field strength is sufficiently high, most of the electrons fail to reach the anode and the electron density is considerably increased, as is the probability of ionizing the gas. For a magnetic field strength of 2.0×10^4 Am^{-1}, the ion current is enhanced by a factor of 2.5×10^4 over its original field-free value and the electron current drops by a factor of fifty. By this means the ratio of ion current to X-ray photocurrent is increased by a factor of 1.25×10^6. Since the cut-off current to the anode is pressure-independent, the ion current to the anode is pressure-independent, and the ion current to the collector just equals the X-ray photocurrent at 3×10^{-14} mbar; this therefore constitutes the low pressure limit of the gauge. Stable gauge operation is only obtained with low emission currents from the filament. This has the additional advantage of reducing gauge pumping. Although this gauge is relatively simple, operates at low (700° C) cathode temperature, has low electronic desorption effects, a low X-ray photocurrent, a linear response and is stable when operated at low emission levels, it does have a major disadvantage; this is its low conversion constant of only 0.06 A mbar^{-1}. In consequence, the low pressure limit is set by the ability of the external current measuring equipment to measure the very small ion collector current, rather than by a residual X-ray current. This gauge is not available commercially.

The low conversion constant cannot be increased by increasing the length of the gauge because of the limited depth of penetration of the electric field produced by the ion collector. Also, the conversion constant cannot be increased by increasing the space charge density in the gauge because of the random production of high-energy electrons that swamp the ion current. These electrons may have equivalent electron temperatures of 10^5 K. These high energy electrons result in instabilities in gauge operation and are common to all types of ionization gauges that employ a high space-charge density with long path lengths.

Lafferty introduced a modified version of this gauge with a high conversion constant for the measurement of extreme ultrahigh vacuum; it retains all the advantages of the original form. In the new design the end plates are retained to prevent the escape of electrons, while the ions are collected by a semicircular segment of a cylinder coaxial with the anode. Another electrode, parallel to this segment, is operated slightly negative or at zero potential. This electrode drains off hot electrons

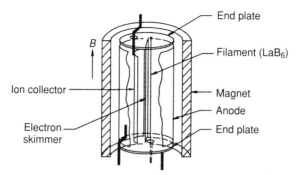

Fig. 5.21 Cut away view of the Lafferty hot cathode magnetron gauge with high conversion constant. The gauge vacuum envelope is omitted for clarity; it would fit just inside the cylindrical permanent magnet.

which might otherwise reach the ion collector. Electrons are supplied by an off-axis LaB_6 coated filament. This new arrangement, Fig. 5.21, provides a gauge with a conversion factor of 0.3 A mbar^{-1} which is linear down to 10^{-14} mbar, and has an X-ray photocurrent pressure limit of 10^{-15} mbar at an emission current of 1 μA. (The limit may really be due to electron-stimulated desorption.)

This completes our review of the principal methods of measuring total pressure. It can be seen that the ionization gauge, in one form or another, is outstanding in its ability to measure pressure over a very wide range; it is consequently the usual choice for low pressure measurements. It is widely available from many different manufacturers who also supply the appropriate electronic control gear. In most cases this control gear contains facilities to operate two types of gauge, e.g. Pirani and ionization gauges, and in some cases three gauges may be catered for, with the inclusion of capacitance manometer control. Generally, this control gear is arranged with computer interfacing as well as outgas facilities. Nevertheless, despite these advantages it must be remembered that ionization gauges do have drawbacks which we may summarize as follows:

1. They require considerable auxiliary electronic equipment.
2. The sensitivity varies for different gases and vapours.
3. The filament may be 'poisoned' in the presence of certain gases.
4. The filament is susceptible to burn-out if exposed to air while hot.
5. Ion bombardment destroys the filament more or less rapidly with continued use.

6. The gauge electrodes must be very thoroughly degassed, especially for use at ultrahigh vacuum.
7. Chemical reactions can occur at the hot filament.
8. Ion gauges pump.
9. They are only accurate to within 25%; UHV gauges may be only accurate within an order of magnitude.

5.15 CALIBRATION FOR DIFFERENT GASES

Penning gauges and ion gauges are usually supplied with control and measuring gear which has, in effect, been calibrated for nitrogen (air). Gas correction factors must be applied if an accurate indication of pressure is required and the gas being measured differs from air. The true pressure is related to the indicated pressure through a gas calibration factor so that

true pressure = meter reading × gas calibration factor

Table 5.2 lists the usual gas calibration factors for a range of common gases. These values must be viewed as approximate.

Table 5.2 Ion gauge calibration factors

Gas	Calibration factor[*]
Air	1.0
Ammonia	1.53
Argon	0.88
Carbon monoxide	0.90
Carbon dioxide	0.71
Neon	2.94
Nitrogen	1.0
Oxygen	0.94
Helium	7.7
Hydrogen	3.1
Methane	0.71

[*] The values given should be regarded as typical. The pressure for a given gas is: true pressure = indicated pressure × gauge calibration factor.

5.16 PARTIAL PRESSURE GAUGES AND RESIDUAL GAS ANALYSERS

All the pressure gauges described thus far measure total pressure. They give no indication of the pressure of individual components, or indeed, what these components might be. Consequently, it may be argued that the only justification these days for using a total pressure gauge on a vacuum system, or particularly a UHV system, is simplicity of operation and cost, since a partial pressure analyser will also identify the residual gases.

The requirements for a high vacuum residual gas analyser and partial pressure gauge are met by the same type of instrument, the term residual gas analyser (RGA) being applied to sophisticated equipment, and partial pressure gauge (PPG) to instruments having a more relaxed specification, particularly with respect to resolution. All high vacuum RGAs and PPGs consist of an ion source, analyser and detector and are distinguished with respect to the method used to separate the ions. The three most important types currently used are the magnetic deflection type, the omegatron and the quadrupole mass filter, with the latter the most convenient and widely available type.

An important parameter of a mass spectrometer is its resolving power. This may be defined in a variety of ways, which in fact lead to different numerical answers. We can define a standard for absolute resolving power by taking the ratio of the peak mass m (measured in atomic mass units, a.m.u.) to the peak width Δm, measured at 10% of the peak height. This is the approach favoured by the American Vacuum Society. Alternatively, we can construct the same ratio now taking Δm as the width at 50% peak height. Of course, analytical mass spectroscopy demands that we discriminate between adjacent mass peaks; this results in two alternative definitions for resolving power. The first, for adjacent peaks of equal height, takes the ratio of m to Δm now measured between the adjacent peaks when the valley height is 10% of the peak height. In the second, for peaks of unequal height, the value of Δm is the peak separation when the main peak makes a 1% contribution to the height of the adjacent, secondary peak.

In quadrupole mass spectrometry, however, it is usual to operate in constant peak-width mode; for example, the full width at 5% peak height is constant (magnetic sector instruments operate at constant resolution $m/\Delta m$ constant). This former mode can result in mass discrimination as sensitivity is traded off against resolution to maintain Δm at a given mass.

Another important characteristic is the sensitivity of the instrument at a given emission current. We define the sensitivity as

$$\text{sensitivity} = \frac{\text{peak signal current}}{\text{partial pressure}} \text{ A mbar}^{-1}$$

The sensitivity is measured by admitting a pure gas to the RGA and measuring the total pressure, here the same as the partial pressure, with a calibrated ionization gauge. Generally, sensitivity is dependent to some degree upon resolution so that increase in resolution leads to a decrease in sensitivity. Sensitivities are universally quoted with respect to the nitrogen peak so that available RGAs are readily compared. In analytical applications of quadrupoles, we are interested in the abundance sensitivity, a measure of the contribution made by a peak at mass m to one at mass $(m-1)$ or $(m+1)$; it is usually measured in ppm and gives a good indication of the quadrupole filter quality.

5.17 MAGNETIC DEFLECTION ANALYSERS

In the magnetic deflection mass spectrometer the gas is ionized by electron impact, with the electrons being supplied by a thermionic filament. The positive ions produced are extracted from an outlet slit of the ion source, accelerated by a p.d. of V volts applied between this slit and a final collimating slit S, before being directed into a uniform magnetic field of flux density B. The magnetic field lines are perpendicular to the ion paths, Fig. 5.22. A positive ion of mass m and charge ne, where e is the electron charge and n is an integer, will undergo a deflection in the magnetic field which results in it following a path along the arc of a circle of radius r given by

$$\frac{mv^2}{r} = Bnev \tag{5.19}$$

where v is the velocity of the positive ion involved. The velocity v is acquired by acceleration through the p.d. V volts, so that we know that

$$Vne = \tfrac{1}{2}mv^2 \tag{5.20}$$

Combining (5.19) and (5.20) yields

$$m = \frac{B^2r^2ne}{2V} \tag{5.21}$$

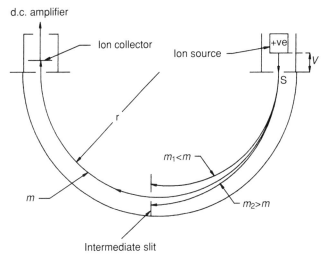

Fig. 5.22 Plan of ion trajectories in a 180° magnetic deflection mass spectrometer. The uniform magnetic flux is perpendicular to the diagram.

Now if a particular value of r is selected by the physical disposition of the entrance and exit slits S_1 and S_2 we know that

$$m = \frac{kne}{V} \qquad (5.22)$$

where k is a constant equal to $\frac{1}{2}B^2r^2$. Usually $n = 1$, so that a scan of values for V is also a scan over the mass spectrum. The atoms or molecules present may be identified from the masses recorded. These instruments are classified according to the deflection angle chosen *viz* 60°, 90°, 120° or 180°. The output from such an instrument will show a series of peaks corresponding to the presence of positive ions of a particular mass to charge ratio, or for $n = 1$, simply the mass spectrum. Further, the relative heights of the peaks will depend on the relative abundance of the various ions present, and with proper calibration, the partial pressure. A typical residual gas spectrum is shown in Fig. 8.1. The most significant factor affecting the peak height of any new mass is the ionization probability within the ion source, and since ionization conditions are similar to those used in ionization gauges, it is possible to correct peak heights. In this way, true total pressures may be obtained as well. In modern instruments gas identification, partial pressure measurement and total pressure measurement is carried out automatically under microprocessor control. In designs which permit of careful outgassing it is possible to read partial pressure of about 10^{-17} mbar.

5.18 THE OMEGATRON

The omegatron introduced by Sommer, Thomas and Hipple in 1951 and adapted for gas analysis by Alpert and Buritz in 1954, may be used for measuring the partial pressures of the various gases in a UHV system. The mode of operation is similar to that of the cyclotron and two views of a simplified version are shown in Fig. 5.23. Electrons emitted from a hot filament are accelerated through the aperature A_1 in the shield box to form a beam which traverses the ionizing region and exits through aperture A_2 before being collected on the anode plate. This beam produces ions in the ionization chamber. A uniform magnetic field of flux density B is maintained by a permanent magnet around the envelope with the field lines parallel to the electron beam passing from A_1 to A_2. A radio-frequency potential difference of 0.5–2 V r.m.s. is maintained across the

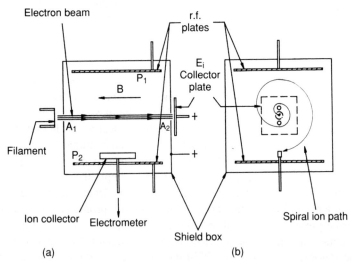

Fig. 5.23 Side (a) and end (b) views of the omegatron, showing electron and ion paths.

electrodes P_1 and P_2 while an ion collector electrode, E_i, is inserted through a slot in either P_1 or P_2. The positive ions produced in the residual gas by the electron beam move towards P_1 during that half cycle when P_1 is negative with respect to P_2. They cannot reach P_1, however, because the magnetic field causes them to follow a curved path of radius r.

For a singly charged ion, r is given by

$$r = mv/Be \qquad (5.23)$$

where v is the mean velocity of the ions. Now the time taken by these ions to describe a semicircle, $\frac{1}{2}T$, is

$$\frac{T}{2} = \frac{\pi r}{v} = \frac{\pi m}{Be} \qquad (5.24)$$

which is independent of r and v. Accordingly, if the frequency of the r.f. signal is adjusted to equal T, then a particular ion of mass to charge ratio m/e will complete its semicircular path just as the r.f. voltage changes sign. The ions are now directed towards P_2 where they are travelling faster, but following a semicircular path of larger radius so that they still complete their path in the same time, $\frac{1}{2}T$. The path of an ion is therefore a spiral, in a plane perpendicular to the magnetic field, which terminates at the ion collector provided that

$$\frac{m}{e} = \frac{BT}{2\pi} = \frac{B}{\omega}$$

ω is the angular frequency of the r.f. signal across plates P_1 and P_2. A frequency scan therefore provides a mass scan at fixed value of B.

The omegatron is compact (the ionizing region may be as small as a 2 cm cube) and readily baked or outgassed. A sensitivity for nitrogen of 10 mbar^{-1} is readily achieved. This is comparable with the hot cathode ionization gauge and a factor of ten bigger than that of the magnetic deflection mass spectrometer. The resolving power is proportional to m^{-1} and therefore decreases with increasing mass of the ion. The omegatron provides an analysis of gases at total pressures in the range 10^{-4}–10^{-11} mbar. Although the omegatron is attractively simple, its performance does not measure up to that of instruments based on the monopole or quadrupole mass filter. As a consequence the omegatron has now largely fallen out of use.

5.19 MONOPOLES, QUADRUPOLES AND ION TRAPS

A very important class of RGAs and PPGs is based on the monopole or, more usually, quadrupole mass filter. The concept of the monopole was introduced in 1963 by von Zahn. Figure 5.24 shows the essential arrangement which is a solid rod set symmetrically in the fold of the right-angled strip or vee block. The separation of the rod from the fold of the strip is r_0. Ions are injected into the space between the rod and

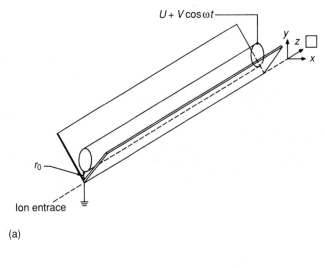

(a)

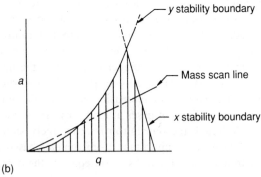

(b)

Fig. 5.24 (a) The structure of the monopole. (b) Ion stability diagram; the shaded area defines stable orbits.

the right-angled strip while a d.c. voltage U plus an r.f. voltage $V\cos\omega t$ is applied to the rod. The ions move in a potential ϕ given by

$$\phi = \frac{\left(x^2 - y^2\right)}{r_0^2}(U + V\cos\,\omega t) \tag{5.26}$$

Virtual images of the rod behind the ground plane of the vee block form the quadrupole field. Ions suffer no change in velocity along the z-axis of the mass filter, but the motion in the x and y directions can be represented by the Mathieu equation,

$$\frac{d^2u}{ds^2} + \left[a_u + 2q_u\cos 2(s - c_1)\right]u = 0 \tag{5.27}$$

Here, $u = x$ or y direction; $s = \omega t/2$ where ω is the angular r.f. frequency; $a_x = -a_y = 8eU/m\omega^2 r_0^2$; $q_x = -q_y = 4eV/m\omega^2 r_0^2$ and c_1 is a constant. There are two classes of solution depending on the values of a and q:

1. The motion remains limited in amplitude as time increases; this is a stable solution.
2. The amplitude becomes infinite after sufficient time; these ions are lost from the device.

Not only must the ions have stable trajectories but they must not strike the vee block; this means that the y-coordinate must always be positive and x always less than y. Ion motion which satisfies these conditions is best seen by examination of Fig. 5.24(b) which shows the boundaries for stability on x and y as a function of a and q. The usual operating condition is shown as the mass scan line. All masses lie on the mass scan line given by $a/q = 2U/V$. The monopole has focussing properties and is sensitive to the ion entrance energy; however, any mass scan line intersects the operating area.

The quadrupole proper consists of four identical rods sets precisely on a square grid; d.c. and r.f. voltages are applied to opposite pairs of rods as depicted in Fig. 5.25(a). Ions are injected at one end and are detected at the other. The stability diagram now looks like Fig. 5.25(b) and the mass line cuts just across the tip of the stability diagram, so that only one e/m species is contained for a given voltage. To scan the mass spectrum, the magnitude of the r.f. potential (or the frequency) is varied while keeping constant the ratio of d.c. to r.f. potentials applied to the electrodes. Species of different e/m ratios are thereby brought one by one into the region of stable paths. This arrangement will accept a wide

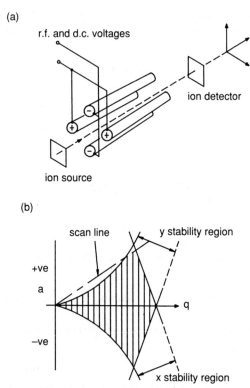

Fig. 5.25 (a) The structure of the quadrupole. (b) Ion stability diagram; the shaded area defines stable orbits.

range of ion velocities in the z-direction. An analogy which helps understand the operation of the quadrupole is that of a ball on a saddle; with the sides of the saddle sloping down and the front and back sloping up, the ball may tend to roll down the sides, but one half cycle of r.f. later the sides slope upwards and the front and back of the saddle slope downwards. With the proper choice of frequency and slope, the ball will oscillate in a complicated manner, but remain in the saddle.

Practical examples of the quadrupole have an ultimate minimum detectable partial pressure of 5×10^{-14} mbar, yet as residual gas analysers they will detect and identify gases at levels of 1–10 ppm. The overall length of the gauge assembly is typically ~260 mm, with a diameter say of 34 mm. They may be operated as partial pressure gauges or total pressure gauges and span the range 10^{-4}–10^{-14} mbar. The control and readout from the modern instruments is fully under microprocessor control with the measurements usually being presented in digital form.

In the monopole and quadrupole, the quadrupole field is in the x–y plane while the ions travel through the spectrometer in the z-direction. However, it is possible to arrange the electrodes to contain or trap the ions, i.e. the ions are stable in all three coordinate directions. The three-dimensional quadrupole field is generated in the structure shown in Fig. 5.26, which is rotationally symmetric about the z-axis. Each electrode is a hyperboloid of revolution. Only one mass species is trapped for a given set of voltage conditions. Ions are detected by pulsing them through the end cap to an ion multiplier located behind the end cap. Gas atoms are ionized inside the spectrometer by an electron beam which traverses the device. The ions formed are stored for appropriate times (depending on the ambient pressure) and are then pulsed out to an electron multiplier. A simple pulse height reading circuit is used to convert the periodic pulses from the electron multiplier to a d.c. signal suitable for driving an X–Y recorder. The ionizing beam is pulsed off while the ion draw-out pulse is applied. The magnitude of the r.f. voltage is scanned while the ratio of r.f. to d.c. voltage remains constant.

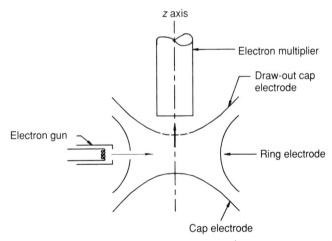

Fig. 5.26 Schematic diagram of electrode arrangement in the three-dimensional quadrupole, or ion trap.

The formation and storage time can be lengthened if the pressure is low, until enough ions of the desired species are present to be detected readily. At 10^{-9} mbar, this would take ~ 25ms. Ions can be stored for more than 60 hours after the ionization process has stopped. The low pressure limit of this ion trap is set only by the storage time employed.

PROBLEMS – CHAPTER 5

5.1 A McLeod gauge is used to determine the pressure in a vacuum system. It has a bulb volume of 350 cm^3 and a capillary diameter of 1 mm. Calculate the system pressure if the mercury column in the capillary is 15 mm from the top.

5.2 A Bayard–Alpert gauge fitted with a modulator electrode records a pressure of 5×10^{-7} mbar when the modulator electrode is at grid potential and 3×10^{-7} mbar when the modulator is grounded. In use under UHV conditions it records a pressure of 2×10^{-10} mbar with the modulator at grid potential and 1.70×10^{-10} mbar with the modulator grounded. What is the true pressure?

5.3 An ion gauge is recording a pressure of 10^{-9} mbar at an ion current of 10 mA. *Estimate* the pumping speed of the gauge at 298 K if 10% of the ions are collected by the gauge envelope and comprise nitrogen ions. The gauge sensitivity for N_2 is 20 mbar^{-1}.

6

Vacuum system design

6.1 REVIEW OF VACUUM SYSTEM COMPONENTS

A complete vacuum system comprises many elements apart from the vacuum chamber itself. Each of these elements should be selected with a view to optimizing performance. We may crudely divide the components of a vacuum system into sub-groups, for example, pumps, gauges, traps and tubing. Pumps in a vacuum system usually operate in a limited pressure range only so that pumping must be performed in several stages, each stage using a different kind of pump; usually no more than two types will be required. Similarly, vacuum gauges have a limited range and we will need to combine different kinds of gauge to span the range from one atmosphere down to the design pressure level; again, we can usually achieve this coverage with two gauges.

The action of operating two different kinds of pump in unison usually results in unwanted interactions between the pumps, which must in turn be controlled by suitably chosen trapping systems. Once again, it is normal to require at least two different types of trapping system.

The operation of pumps, gauges and traps can only occur if they are suitably connected together, with due regard to the likely pressure regimes which will exist in each part of the pumping system. It will normally be the case that the overall system performance will be controlled by the dimensions of the tube connecting the vacuum chamber to the first pump. The dimensions of this tube must be carefully considered in order to meet the design criteria. These may impose limits on pump-down time, final pressure or gas throughput. For the ultrahigh vacuum range, further special considerations must be made so that the virtual leakages from all sources may be minimized or eliminated.

In this chapter we shall examine the way the selection and balancing of components may be achieved for a given performance objective. We shall try to quantify the various elements of a pumping system's performance and how they are matched in practice.

6.2 GAS FLOW BETWEEN CHAMBER AND PUMP

Very roughly there are two kinds of vacuum system we must consider:

1. Those in which we wish to maintain a very high vacuum by continuous pumping.
2. Those in which the system will eventually be sealed off and disconnected from the pumping line.

These two distinct uses actually impose important restrictions on the nature of the connection between the system and the high vacuum pump. In our first group we must insist that the conductance, F_p, of the tube or pipe joining the vacuum pump to the system be much larger than the intrinsic speed of the pump we wish to use. Equation (2.38) relates the observed speed, S'_p, at the vacuum chamber to the intrinsic speed of the pump, S_p, and the connecting conductance, F_p. It is convenient to rewrite this expression in a different form which gives the ratio S_p/S'_p as a function of the ratio $F_p/S_p = n'$, thus

$$\frac{S_p}{S'_p} = \frac{n' + 1}{n'} \tag{6.1}$$

Figure 6.1 plots the ratio S_p/S'_p against n' and shows therefore the effect of the conductance F_p. In this case S_p must also be large enough to cope with the expected gas load.

In our second group, the conductance of the pipe joining the vacuum pump to the system is likely to be very low and the value of n' very small. Under these circumstances, it is the nature of the connecting tubing which controls the performance of the vacuum system and large values for S_p are clearly of little advantage.

To summarize; our continuously pumped system requires $F_p \gg S_p$ and $S_p > Q_T/p_u$, where Q_T is the total gas load of the system from all sources and p_u is the ultimate pressure attainable. In our sealable system $F_p \ll S_p$, and there is often no special restriction on S_p since it is the behaviour of the connecting tube which dominates the system behaviour.

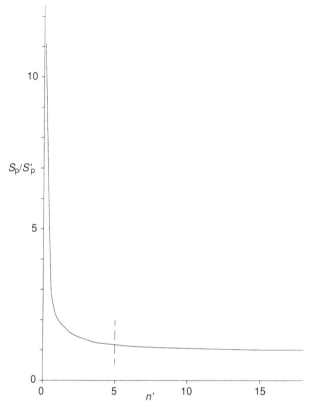

Fig. 6.1 Plot of the ratio S_p/S'_p against n', where S_p is the intrinsic pumping speed of the pump and S'_p the actual pumping speed at the vacuum chamber mouth; n' is the connecting pipe conductance divided by the intrinsic pump speed, i.e. F_p/S_p. In practical applications, n' should be at least equal to five.

It is convenient to begin our analysis of the problems associated with a real vacuum system by considering just the vacuum chamber and then adding the parts required for an operating system. We have already defined the speed of a pump, S_p, in section 2.8, but it is worth emphasizing again that, although pump speed has the dimensions of conductance, i.e. ls^{-1} or m^3s^{-1} and may be used like conductance, it is in fact the throughput divided by the pressure in a specified plane, rather than the throughput divided by the pressure drop across a pipe. Thus, in Fig. 6.2 we define the pumping speed in the plane of the outlet pipe as S'_p which in turn equals Q_T/p_0 or dV/dt; Q_T is the sum of the various gas loads.

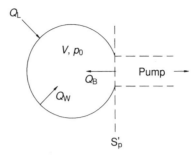

Fig. 6.2 Schematic diagram of a vacuum chamber showing the gas flows which normally exist. The chamber has a volume V and operates at a pressure p_0; gas throughputs are represented by Q_L (leakage or permeation), Q_W (outgassing from walls) and Q_B (backstreaming from the pump). Again, S'_p defines the actual speed in the mouth of the vacuum chamber.

6.3 RESIDUAL GAS SOURCES

Figure 6.2 defines the individual gas loads present in a vacuum system of volume V. In effect, there are always three gas sources: gas backstreaming from the pump, Q_B, gas desorbing from the walls, Q_W, and gas leaking in, Q_L. In this last category we can include gas permeation through the walls, although this would not manifest itself unless the pressure was reduced to the UHV range and beyond. All these processes are assumed to be taking place in an isothermal system. From Boyle's law, the product pV must be a constant at constant temperature during the volume change dV. Therefore,

$$d(pV)/dt = 0$$

and we can write

$$p \frac{dV}{dt} + V \frac{dp}{dt} = 0 \tag{6.2}$$

Substituting our definition of S'_p we obtain

$$p\, S'_p\, dt = -V dp \tag{6.3}$$

The minus sign is, of course, required since dp describes a pressure drop. For our idealized system in Fig. 6.2, we must rewrite (6.3) to include the additional gas sources, so that

$$-V dp = dt(S'_p\, p - Q_W - Q_L - Q_B) \tag{6.4}$$

6.4 ULTIMATE PRESSURE, SPEED OF EXHAUST

After the initial gas removal, it is the auxiliary gas sources represented by Q_W, Q_L and Q_B which determine what gas is left in the system. It was pointed out in Chapter 3 that special measures may well be necessary to control or eliminate gas inputs from these sources: in any event it is the gas from these sources which will control the final pressure. Thus, we can say that the ultimate or equilibrium pressure attainable, p_u, will be reached when dp/dt is zero, or

$$p_u S_p' = Q_W + Q_L + Q_B \tag{6.5}$$

and

$$S_p' = \frac{Q_T}{p_u} \tag{6.6}$$

We can define an effective pumping speed S_e which takes account of these additional gas sources so that (6.5) becomes

$$S_e p = S_p' p - Q_T = S_p' \, (p - p_u) \tag{6.7}$$

and

$$S_e = S_p'\left(1 - \frac{p_u}{p}\right) \tag{6.8}$$

Note that S_e is a function of pressure and becomes zero at the ultimate pressure. It is also known as the speed of exhaust. In general use, we can control Q_W and Q_L by proper design or proper processing, so that they may be ignored: Q_B will then determine p_u.

The gas source Q_L is usually constant, whether it be constituted from a real leak or simply permeation; Q_B and Q_W change only slowly with time. If we assume that, in fact, Q_B, Q_L and Q_W are constant and that S_p' is constant, then we can integrate (6.4) to obtain

$$p - p_u = (p_0 - p_u)\exp\left(\frac{-S_p' t}{V}\right) \tag{6.9}$$

where p_0 is the pressure at $t = 0$. Now p_0 is much larger than p_u so that we can write

$$p = p_0 \, \exp\left(\frac{-S_p' t}{V}\right) + p_u \tag{6.10}$$

In (6.9), V/S'_p has the units of time, which we can write as $V/S'_p = \tau$. Essentially τ is the system time constant and a useful index for judging vacuum system performance. It should always be known for a given vacuum system.

6.5 PUMP-DOWN TIME

The foregoing analysis deals with the behaviour of a vacuum system in which we have specified the pumping speed in the plane of the outlet orifice. In a real system of course this point would be the opening of a pipe connecting the vacuum system to a vacuum pump and we need to know how this combined system will behave. The vacuum pump has an intrinsic speed S_p, stated by the manufacturers, which is that which is obtained in a plane at the entrance to the pump. The problem is to determine how the interconnecting pipe influences the pumping behaviour of the system. If a pump of speed S_p is connected to our system via a pipe of conductance F_p (Fig. 6.3), we know from (2.38) that the pumping speed of the combination, S'_p, is just $S_p F_p/(S_p + F_p)$. Clearly F_p must be at least equal to S'_p but preferably much larger. The problem of determining the vacuum system behaviour with time now largely resolves into one of taking account of the behaviour of F_p with pressure and hence time.

If we wish to know the pump-down time from a pressure p_1 at time t_1 to a pressure p_2 at time t_2, we simply return to (6.4) and integrate between these limits, thus,

$$t_2 - t_1 = \frac{V}{S'_p} \ln\left(\frac{p_1}{p_2}\right) \tag{6.11}$$

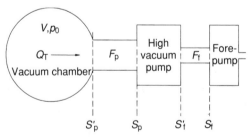

Fig. 6.3 Schematic diagram of a typical vacuum system defining the pumping speeds at the inlet of the fore pump. The total gas load from the chamber of volume V is Q_T, p_0 is the operating pressure, and F_p and F_f are the connecting pipe conductances.

where S'_p is the speed at the chamber mouth, assumed constant. It has been pointed out earlier, in section 6.1, that it is not usually possible to operate a vacuum system with a single pump, but that usually two will be required and very occasionally three. Under these conditions we can replace (6.11) with

$$t_2 - t_1 = V\left[\frac{1}{S'_p}\ln\left(\frac{p_2}{p_f}\right) + \frac{1}{S'_f}\ln\left(\frac{p_f}{p_1}\right)\right] \tag{6.12}$$

where p_f is the intermediate pressure achieved using the fore pump alone, and S'_p is the speed at the entrance to the fore pump tubulation, i.e. the exit from the high vacuum pump.

If we are interested in the second kind of system where by necessity the conductance of the connecting tubing dominates performance, we have to take account of the variation in the tube conductance with pressure and hence time, since F_p is not constant. The variation in F_p as we go from viscous to molecular flow is most simply handled by means of Knudsen's approach embodied in (2.31) and (2.32). If we transform (2.31) to a more convenient form we get

$$F = F_M\left(\frac{F_V}{F_M} + X\right) \tag{6.13}$$

We can substitute for the components of F_V and F_M to obtain

$$F = F_M\left(\frac{0.147a}{\lambda_a} + X\right) \tag{6.14}$$

where λ_a is as defined for (2.32) and a is the tube radius.

Let us assume that the whole of the pressure drop between our system of volume V and the vacuum pump inlet occurs across the connecting tube, i.e. the pump inlet pressure is effectively zero. Therefore p, the pressure at the entrance to the tube, is given by

$$p = 2p_a$$

We will also write

$$\lambda_a = \lambda_1/p$$

where λ_1 is the mean free path at 1 mbar and p is in mbar. Equation (6.3) may be rewritten as

$$pF\mathrm{d}t = -V\mathrm{d}p \tag{6.15}$$

Substituting for

$$F = F_M \left(0.0736 \frac{a}{\lambda_1} p + 1 \right)$$

we get

$$-\frac{dp}{dt} = bp^2 + kp \tag{6.16}$$

where b and k are constants; $b = 0.0736a \, F_M / \lambda_1 V$ and $k = F_M / V$. The value of X may, with little error, be assumed equal to unity. Equation (6.16) transforms to

$$\frac{-dp}{p} + \frac{dp}{(p + k/b)} = k dt \tag{6.17}$$

and integration yields

$$-\ln p + \ln\left(p + \frac{k}{b} \right) = kt + \text{const} \tag{6.18}$$

If we integrate between the limits of $p = p_1$ at $t = t_1$ and p_2 at $t = t_2$ where $t_2 > t_1$ and $p_1 > p_2$, then

$$\ln\left(\frac{p_1}{p_2} \right) - \ln\left(\frac{p_1 + k/b}{p_2 + k/b} \right) = k(t_2 - t_1)$$

Equation (6.19) does not look very appealing or convenient to use, but we can achieve simplification and physical insight if we consider two limiting cases, namely, when $p_2 > 10$ Pa or 0.1 mbar, then $p_2 \gg k/b$. Here we can make use of the fact that the left hand side of (6.18) then reduces to k/bp, so that (6.19) becomes

$$\frac{1}{b}\left(\frac{1}{p_2} - \frac{1}{p_1} \right) = t_2 - t_1 \tag{6.20}$$

and $1/p$ varies linearly with t; this is the viscous flow regime.

In the molecular flow regime $p \ll k/b$ and $p_1 < 0.01$ mbar or 0.1 Pa. Equation (6.19) then becomes, as before

$$\ln\left(\frac{p_1}{p_2} \right) = k(t_2 - t_1) \tag{6.21}$$

The time required to pump down from p_1 to p_2 is given by substituting for k, thus

$$t_2 - t_1 = \frac{V}{F_M} \ln\left(\frac{p_1}{p_2}\right) = 2.303 \frac{V}{F_M} \log\left(\frac{p_1}{p_2}\right) \qquad (6.22)$$

For a vacuum system designed in such a manner that $F_p \gg S_p$ and $S'_p \approx S_p$ during pump-down, because of the high viscous flow conductance, we can see that we can return to (6.11) simply by replacing F_M by S'_p. When S'_p is independent of p, a plot of $\ln p$ against t yields a straight line of slope $S_p/2.303V$, so that if V is known, S'_p may be determined.

6.6 DESIGN OF A HIGH VACUUM SYSTEM

In order to clarify the ideas presented previously, it is useful to consider the design of a vacuum system of the first kind, that is, one intended to operate continuously and provide a stated pressure against a given gas load. The actual components involved will not change if the system is required for intermittent use, although the system performance will then be dominated by the characteristics of the connecting tubes.

A high vacuum system of the first kind will generally consist of the following basic elements:

1. The vacuum chamber of volume V which it is required to evacuate to a pressure p_0.
2. A connecting pipe of conductance F_p joining the vacuum chamber to the high vacuum pump, which will generally be a diffusion pump.
3. A high vacuum pump which has speed S_p at the pressure p_0 and which is capable of operating against a backing pressure no higher than p_f.
4. A connecting pipe of conductance F_f joining the high vacuum pump to the fore-pump or backing pump.
5. A fore pump of speed S_f at fore pressure p_f.

Figure 6.3 shows this system schematically. We must also include a term for the gas load arising, either from gas introduced into the system, or from gas desorbing from the vacuum chamber surface. Let us assume that this is Q_T.

Generally, the design of a vacuum system will begin from performance specifications such as the pump-down time t, maximum gas throughput Q_T, and chamber operating pressure p_0. In addition, the

chamber volume V will be known. Sometimes there will be special requirements which stipulate, for example, no oil vapour or an operating
temperature limit. The former will dictate the choice of pump type. We
begin by calculating the pumping speed required at the vacuum
chamber S'_p, to attain the operating pressure with the specified gas load.
We know that

$$S'_p = \frac{Q_T}{P_0} \qquad (6.23)$$

Also

$$\frac{1}{S'_p} = \frac{1}{S_p} + \frac{1}{F_p} \qquad (6.24)$$

and

$$S_p = \left(\frac{n'+1}{n'} \right) \frac{Q_T}{P_0} = f \frac{Q_T}{P_0} \qquad (6.25)$$

where $n' = F_p/S_p$ is plotted against the ratio S_p/S'_p in Fig. 6.1. In
designing a vacuum system, it is usual practice to choose values of F_p
which will provide a large value for n'. Inspection of Fig. 6.1 shows
that if $n' = 5$ say, then S_p/S'_p is 1.2; or if $n' = 10$, $S_p/S'_p = 1.1$. A value of
n' between 5 and 10 therefore yields an actual chamber speed close to
the pump speed and is good design practice. Similar arguments apply to
the tubulation for the fore pump.

The characteristics of the fore pump are determined by, at most, two
requirements, the required pump-down time and, in the case of diffusion pumps or vapour ejector pumps, the minimum backing pressure
necessary. The minimum backing pressure is stated for all diffusion and
vapour ejector pumps as a function of pump fluid. To determine the
rotary pump speed required to meet the critical backing pressure
requirement is simply a matter of determining the maximum throughput
for the diffusion or vapour ejector pump. This is obtained from the
manufacturer's speed pressure curves and is the product of pressure and
speed, at 10^{-2} mbar for diffusion pumps and at 1 mbar for vapour
ejector pumps. The throughput divided by the critical backing pressure
yields the minimum backing pump speed required.

The value obtained for the rotary pump speed, although meeting the
diffusion pump requirements, may not be sufficient to pump the system
down in the stated time t. The pump-down time t is determined using
(6.11) with the appropriate values for the pressures; thus we may write

$$t = 2.303 \frac{V}{S_f'} \log\left(\frac{1013.2}{p_f}\right) \tag{6.26}$$

For most situations, a value for p_f of 10^{-1}–10^2 mbar is appropriate. Equation (6.26) provides a value for S_f' and thus F_f may be determined by setting an appropriate value for n' again; this yields S_f. Of course, the values obtained for F_p and F_f must embrace the conductance of any cold traps, sorption traps or valves which form part of the fore line and pumping line.

The basis for the calculation of pump-down time is the assumption that **the pumping speed is constant**. While this is generally true for rotary pumps, it is not true for diffusion or turbomolecular pumps as inspection of Fig. 4.13(b) will show. Consequently, when a pressure is reached at which the throughput of the rotary pump equals that of the diffusion or turbomolecular pump, the latter pumps are brought into action and now the time required to reach maximum speed is calculated on the basis that **throughput is constant**. The time required for this stage of the pump down process is given by

$$t \simeq \frac{V}{S_p'}\left(\frac{p_e}{p_m} - 1\right) \tag{6.27}$$

where p_e is the pressure at equality of throughput and p_m is the pressure at which maximum pumping speed is attained; around 10^{-3} mbar for most diffusion pumps. The time involved in this part of the pump-down process is usually only a fraction of a minute and can often be ignored. Below 10^{-3} mbar, the main vacuum pumps, diffusion, turbomolecular and cryopumps have constant speed and the equilibrium pressure is determined by the gas load, $Q_T = S_p' \times p_0$.

6.7 BACKING PUMPS

Reference to Fig. 4.1 shows that the number of pumps capable of operating from atmospheric pressure is very limited. In fact, although four pump types are apparently capable of doing so, only three, the sorption pump, the hook and claws pump and the rotary oil pump, are used this way; they constitute backing pumps. The choice between them is straightforward insofar as the sorption pump has limited pumping capacity but produces no contamination, whilst the rotary oil pump and the hook and claws pump have unlimited pumping capacity (ability to handle continuous gas loads). However, the hook and claws pump operates dry and unlubricated, and produces no significant contamination load at its input; on the other hand, the rotary oil pump (either rotating vane or eccentric piston) produces a significant quantity of oil cracking products at its inlet which must be removed by trapping. As mentioned previously, the way to do this is by means of an activated alumina or molecular sieve trap, with the usual choice being activated alumina run at room temperature. The typical output of hydrocarbons at the inlet to a rotary oil pump, and the effect of an activated alumina trap on this output, is shown in Fig. 6.4. The hook and claws pump finds special use in the semiconductor processing field where the continuous pumping capability and the absence of oil vapour is an especial advantage.

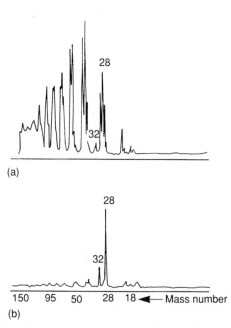

Fig. 6.4 (a) Residual gas spectrum at the inlet of a rotary oil pump showing the cracking products and residual hydrocarbons which, if not trapped, will backstream through the high vacuum pump to the vacuum chamber. (b) Residual gas spectrum after insertion of an activated alumina trap showing almost complete removal of hydrocarbon fractions leaving N_2 (28) and O_2 (32).

In those situations where higher speeds or lower ultimate pressures are required than those available for rotary oil pumps or hook and claws pump alone, use is made of the Roots pump operating in series to improve performance. The Roots pump generally has a specified maximum pressure difference between the inlet and outlet in order to prevent overheating of the rotors and consequent pump seizure. This pressure difference is commonly of the order of 10 mbar but this value may be exceeded for short operating times. It will therefore not usually be used for pumping from atmospheric pressure directly. The Roots pump is used in one of two modes, either as a transport pump or as a compression pump.

In the transport mode it is backed by a rotary oil pump of the same pumping speed, with both pumps being started simultaneously at atmospheric pressure; (the critical pressure drop will not be exceeded under these conditions). This arrangement permits achievement of a lower ultimate pressure than that available from the rotary pump alone.

In the compression mode, a Roots pump is placed in series with a rotary pump having a speed 5–10 times smaller than that of the Roots pump. Initially, at atmospheric pressure, the Roots pump is bypassed and all the pumping is done by the rotary pump until the pressure difference across the Roots pump falls below the critical value. At this point in time, the bypass is closed and the Roots pump participates in the pumping process. This combination affords a lower ultimate pressure and a higher overall pumping speed. Very large turbomolecular pumps or diffusion pumps are generally backed this way; combined Roots pumps and rotary oil or hook and claws pumps are available from manufacturers.

The sorption pump, which operates with molecular sieve material chilled to 77 K, is the appropriate choice for at least three of the high vacuum pumps in current use. More than one sorption pump should be used, operated in series, each one being chilled down in succession and then valved off. Their ability to deal with the rare gases is limited, but

Table 6.1 Properties of principal backing pumps

Backing pump type	Pumping capacity	Background contamination
Sorption pump	limited	none
Hook and claws	unlimited	none
Rotary oil pump	unlimited	hydrocarbons

this is not usually important since the high vacuum pump will remove them. The properties of the principal fore pumps are summarized in Table 6.1.

6.8 HIGH VACUUM PUMPS

There are basically five pumps to choose from when selecting a high vacuum pump. The choice depends entirely on the type of system which is being evacuated. A system operating with a permanent gas flow, such as a molecular beam system, would be pumped by a diffusion pump or turbomolecular pump by choice, though this could be augmented with a cryopump, since in principle, the pumped gas may simply be stored as frozen layers on the cryopump panels. Systems operating without a continuous gas load would be pumped using a pump chosen to fit the particular vacuum requirements. Thus, if there is a particular need to operate with no organic contamination, the choice would be made from the turbomolecular pump, getter-ion pump, getter pump or cryopump.

Development of the turbomolecular pump over the last decade has meant that this pump is often the first choice for a system requiring freedom from organic contamination; it has the additional advantage of being able to cope with continuous gas flows. Perhaps the most important decision with regard to diffusion pumps is the selection of an appropriate pumping fluid. The choice will normally lie between Santovac 5, a polyphenyl ether, if UHV work is contemplated, or one of the cheaper silicone oils if more general high vacuum work is contemplated. The choice of pump fluid is also relevant to the selection of a backing pump. In the vapour stream from the top nozzle of a diffusion pump jet system, the pump fluid molecules not only travel towards the cold walls of the pump, but also receive momentum in collisions which results in them moving out of the jet flow and towards the vacuum chamber. This phenomenon is known as backstreaming and will normally amount to no more than a few micrograms per minute for every cm^2 of pump inlet area; nevertheless, this must be dealt with in more demanding applications.

The solution to the backstreaming problem is to introduce a water-cooled baffle which will trap as many backstreaming pump fluid molecules as possible while offering high conductance. These two requirements are mutually exclusive so all designs are a compromise.

An alternative to a baffle is a cold trap. Like baffles, cold traps prevent backstreaming of pump fluid molecules and are so designed that back-diffusing molecules are caused to impinge more frequently on cold surface deflectors than would be the case with a baffle. Cold traps are vital if one is using a diffusion pump based on mercury, or where it is necessary to obtain a vacuum free of oil vapour for long periods. Cold traps are operated at 77 K, liquid nitrogen temperature. Baffles and cold traps are given additional consideration in section 7.9.

6.9 ULTRAHIGH VACUUM

To obtain an ultrahigh vacuum, that is pressures less than 10^{-8} mbar or 10^{-6} Pa, requires consideration of many factors which would be thought trivial in high vacuum work. These factors determine the quantities Q_L, Q_W and Q_B described earlier in this chapter. Now we know from (6.24) that the equilibrium pressure p_u in the high vacuum chamber, if the effective pumping speed is S'_p, will be given by

$$p_u = \frac{Q_B + Q_L + Q_W}{S'_p} \tag{6.28}$$

so that p_u can be reduced either by decreasing Q_B, Q_L and Q_W or by increasing S'_p. The production of ultrahigh vacuum is then simply a matter of reducing the ratio of the gas influx to the pumping speed. The various elements of this gas influx are sketched in Fig. 6.5 which shows the usual time scale for pressure reduction in a vacuum system. The vacuum production process divides into four time zones, each time zone being largely controlled by a particular gas source which constitutes the rate-limiting process. The initial and most important part of the pressure reduction process is the volume removal of gas; this process has an exponential dependence on time, (6.10), and is completed relatively quickly if the correct pumps have been selected. The time scale indicated in Fig. 6.5 is arbitrary for this part of the evacuation process and could easily take longer than shown.

Once the volume removal of gas is complete, the vacuum is controlled by gas desorbing from the vacuum chamber surfaces; this process shows a t^{-1} dependency and is the major contributor to p_u in (6.10). It is worth noting that the time scale for removal of all significant amounts of absorbed gas is very long at room temperature, before the gas load starts to be controlled by gases diffusing into the vacuum chamber from the component elements: this process shows a $t^{-1/2}$ dependence. Finally, the gas load is from permeation, that is, gas adsorbing on the outside of the chamber, diffusing through the walls

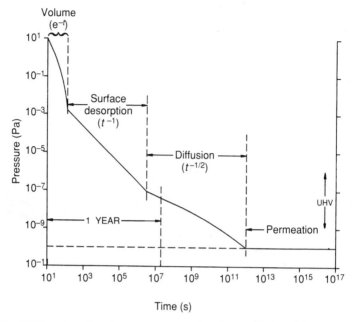

Fig. 6.5 Log–log plot of pressure versus time for a typical, unbaked vacuum system, showing the rate-limiting processes which determine the ultimate vacuum attainable. The principal component of the gas phase in the surface desorption region is likely to be water: in the lower pressure regimes it is likely to be hydrogen.

and desorbing from the inside surfaces: this process is simply temperature dependent.

With the exception of the volume removal of gas, all of these processes are very dependent on temperature but, at room temperature anyway, the pressure decrease takes place so slowly that the diffusion and permeation controlled portion of the gas-release process cannot be reached on a reasonable time-scale. High temperature processing is necessary to bring the time-scale down to a reasonable value. Recognition of the need to do this existed quite early on in the development of vacuum technology (1930s–1940s), but it was the introduction of the Bayard–Alpert inverted ion gauge in the 1950s which provided the final step: the means with which to measure ultrahigh vacuum. Without reliable pressure measuring systems, the early workers could only infer, by indirect means, that the pressure was below 10^{-6} Pa or 10^{-8} mbar.

The quantity Q_B is the gas backstreaming from the pump. This is generally more of a problem with diffusion than with other types of high vacuum pump, but as has previously been mentioned, it can be reduced by careful first jet design (cooled guard rings) and appropriate choice of pump oils. The problem exists for the other types of high vacuum pump as well, but takes a different form. Thus, although the getter-ion pump is free of significant hydrocarbon contamination, it does release hydrogen, methane and noble gases, whilst the turbomolecular pump tends to release principally hydrogen, but also water and carbon dioxide. It will release hydrocarbons as well if it has conventionally lubricated bearings. Only the magnetic levitation bearing systems avoid this problem. Any high vacuum pump which is backed by an untrapped rotary oil pump will backstream hydrocarbons from the fore-pump, so activated alumina fore-line traps are mandatory. Conventional liquid nitrogen cold traps inserted between the vacuum chamber and the high vacuum pump can reduce the presence of hydrocarbons and water vapour, but they also act as gas sources. In practice (6.28) mis-states the situation and has to be written as

$$p_u = \frac{Q_B + Q_L + Q_W}{S'_p} + p_r \qquad (6.29)$$

where an additional term p_r, the residual pressure, has been inserted. This term arises because the trapping efficiency of a cold trap is not 100% and the backstreaming from a diffusion pump, or any other high vacuum pump, increases with pump size. Consequently, if the pump size and the cold trap surface are increased, the backstreamed gas load and the pumping speed go up in proportion leaving a residual pressure affected only by the trapping efficiency. If the cold trap is omitted there is still a base pressure since, for example, more hydrogen backstreams from a large ion pump than a small one.

Equation (6.29) tells us that, although one might be tempted to try to achieve UHV simply by increasing pumping speed, the most efficient way, in fact, is to reduce the total gas load Q_T since increasing S'_p will not decrease p_r.

Figure 6.5 shows that the most important barrier to the attainment of UHV is gas desorbing from the walls and materials of the vacuum chamber; this gas is represented by Q_W. Gases chemisorbed on the metal components of the vacuum chamber or adsorbed on the glass surfaces of viewing ports, as well as the vapour pressure exerted by construction materials, contribute to this total. Chemisorbed and other absorbed material, usually water in the case of glass surfaces, may be largely removed by baking the vacuum chamber at elevated temperatures, up to 250° C for stainless steel or 450° C for glass. The need for high temperature bakeout to remove chemisorbed gases places restrictions on the other materials which may be used in construction, particularly vacuum seals.

In general, vacuum seals fall into two classes, static or dynamic (motion) seals. The first category may be further subdivided into permanent and demountable seals. Permanent static seals would comprise glass-to-metal seals in a glass ultrahigh vacuum system, or welded joints in a stainless steel system. While glass-to-glass or glass-to-metal seals are straightforward, welded joints require careful preparation and execution since they must be made as far as possible entirely within the vacuum chamber to avoid the production of voids and narrow fissures which will entrain gases.

Demountable seals include rubber gaskets and O-rings, or metal gaskets which depend on elastic or plastic deformation of soft metals such as copper, gold, aluminium or indium. The former, rubber seals, were commonly used on high vacuum systems, but rubber itself is gassy, exuding H_2O, CO, N_2 and CO_2 as well as a variety of hydrocarbons for long periods of time, and additionally, is quite permeable to atmospheric gases and of course totally unstable at the temperatures required for bakeout. Although natural rubber and synthetic derivatives of it are unsuitable for ultrahigh vacuum work, the new fluroelastomers marketed under the name of Viton® or Kalrez® do not suffer from these problems and may, in the case of Kalrez® anyway, be baked to 250° C. They permit the achievement of vacua in the 10^{-8}–10^{-9} mbar (10^{-6}–10^{-7} Pa) range.

If a metal seal is to be used in ultrahigh vacuum work it should enable continuous molecular contact to exist between the mating surfaces. This is achieved by plastic deformation of metals, mating of smooth surfaces by squeezing, or by elastic deformation. The details of these metal seals, both static and dynamic, will be described in Chapter 7.

The final gas source comes under the label Q_L, where Q_L may be a real leak (this normally precludes UHV), or simply the effect of permeation through the fabric of the vacuum chamber. There are some basic limitations which prevail on the value of Q_L, so that at pressures below 10^{-10} mbar (10^{-8} Pa), atmospheric helium will, for example, permeate through the walls of a glass pyrex vacuum system at a rate which is measurable in a sealed-off system. Similarly, hydrogen permeates most metals, particularly iron, and will provide a limit for a stainless steel system. All of the refractory metals and noble metals are suitable for vacuum use, as are the stainless steels, and oxygen-free high conductivity copper (OFHC). The choice of a metal for use in an ultrahigh vacuum system should always pay attention to its immediate environment. Thus, metal components which are likely to run hot must always be capable of being outgassed at a temperature higher than that at which they will normally operate. A summary of the properties of materials used in vacuum work may be found in Appendix B. The problem of real leaks will be dealt with in Chapter 8.

The preceding account has pointed up the dominant effect of surface processes in the attainment of UHV. It must be emphasized that the special feature which distinguishes UHV systems from ordinary systems is cleanness, that is, the elimination of contamination from all sources. The outgassing loads may be reduced by meticulous cleaning processes for all components of the vacuum chamber. Ideally all stainless steel components will be electropolished at the final stage of manufacture and vapour degreased in perchlorethane before use. No parts of the vacuum chamber may be touched by the ungloved hand since fingerprints represent a gas source on their own. Further, component parts should not be left in contact with rubber bands or plastic bags since they can readily contaminate with organic residues. Clean (degreased) aluminium foil is a satisfactory alternative.

Figure 6.6 shows a schematic diagram of the arrangement of components to form a basic stainless steel system capable of attaining UHV. If the oven facility is dispensed with it will be capable of good high vacuum performance. The vacuum chamber pressure is measured with a nude Bayard–Alpert gauge, or better still a quadrupole mass spectrometer, or both, and trapped by a liquid nitrogen cooled cold trap, all of which items must be baked; the ion gauge can be outgassed at red heat. The chosen high vacuum pump will lie outside the bakeout zone if it is a diffusion pump, but may well be partially baked out during operation if it is a getter-ion, getter or turbomolecular pump. Particularly in the latter case, the bakeout temperature will be a good deal less than 250° C, usually 80–100° C, and in view of the large surface area presented by the fixed and rotating vanes, vital to the attainment of UHV. The fore-pump or backing pump section is mainly comprised of a pump chosen to suit the properties of the high vacuum pump (Table 6.2). An activated alumina trap must be included if the backing pump is any form of rotary oil pump. This trap should be capable of being isolated from the fore-pump and the high vacuum pump by means of isolating valves and it serves also to protect the Pirani gauge from contamination by decomposition products, etc., from the fore-pump. It is usually convenient to have at this point an inlet valve to allow the

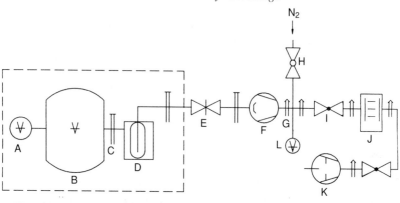

Fig. 6.6 Schematic diagram of a basic stainless steel UHV system using standard vacuum symbols. Here, A is the Bayard–Alpert gauge or quadrupole mass spectrometer head; B is the vacuum chamber; C is a bolted flange coupling (Conflat); D is the liquid nitrogen trap; E is the gate valve; F is the diffusion pump; G is the KF coupling; H is the dry nitrogen admittance valve; I is the straight through valve; J is the activated alumina trap; K is the rotary oil pump and L is the Pirani gauge. The region enclosed by the broken line must be baked at up to 200° C if UHV is to be obtained, otherwise only high vacuum will be reached.

system up to atmospheric pressure. Best practice is to use dry nitrogen for this purpose, maintaining a slight nitrogen overpressure even when at atmospheric pressure, with the arrangement shown, there will be little tendency for rotary pump fractions to be swept into the vacuum

Table 6.2 Properties of high vacuum pumps

Pump type	Pumping capacity	Principal background vacuum components	Preferred backing pump
Diffusion pump	unlimited	H_2, CH_4, C_2, H_4, higher hydrocarbons	Rotary oil, Hook and claws
Turbomolecular pump	unlimited	H_2, He,	Rotary oil, Hook and claws
Getter-ion pump	limited	H_2, CH_4, memory effects	Sorption
Getter pump	limited	H_2, CH_4, memory effects	Sorption
Cryopump	limited	He	Any

chamber, although the high vacuum pump fluid vapours may be so transported. It is envisaged that the whole system will be pumped down initially by just the backing pump, until the Pirani gauge indicates that the pressure is low enough for the high vacuum pump to begin operations. If the high vacuum pump is a turbomolecular pump it will generally be started simultaneously with the backing pump.

If the system is constructed from glass then it will be as shown in Fig. 6.7, where the only differences are the absence of valves and

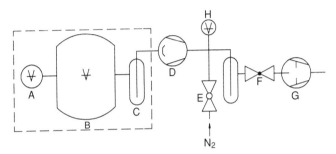

Fig. 6.7 Schematic diagram of a basic glass UHV system using standard vacuum symbols. Here, A is the Bayard–Alpert gauge; B is the vacuum chamber; C is the cold trap; D is the diffusion pump; E is the dry nitrogen admittance valve; F is the straight through valve; G is the rotary oil pump and H is the Pirani gauge. The region enclosed by the broken line must be baked at up to 450° C if UHV is to be obtained. Note the general absence of valves and joints.

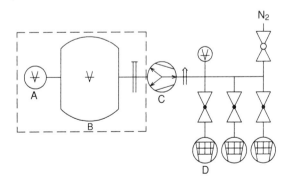

Fig. 6.8 Schematic diagram of a stainless steel vacuum system based on a getter-ion pump; A is the Bayard–Alpert gauge; B is the vacuum chamber; C is the getter-ion pump and D are the sorption pumps.

flanges together with the introduction of a liquid nitrogen-filled cold trap as the means of trapping oil breakdown products. In this case best practice would involve a gas inlet tube, which is drawn down to a capillary and sealed off. If it is necessary to bring the system up to atmospheric pressure this is done by scribing a groove in the capillary section, slipping the nitrogen inlet tube over it and then cracking off the capillary. This procedure avoids the use of a greased stopcock.

Figure 6.8 shows, using the standard symbols, the arrangement required if the vacuum chamber is pumped by a getter-ion pump backed by three sorption pumps. This sort of arrangement is suitable only for a large stainless steel vacuum chamber; fewer sorption pumps are necessary for a small vacuum chamber.

A listing of the standard vacuum symbols may be found in Appendix A.

PROBLEMS – CHAPTER 6

6.1 A vacuum chamber has a volume of 100 litres and an operating gas load of 10^{-4} mbar ls^{-1}. The design operating pressure is 10^{-7} mbar. Connections between the chamber and diffusion pump and the diffusion pump and rotary pump are to meet good design practice with the value of n' chosen as 5. Calculate the pumping speed at the chamber, the minimum connecting pipe conductance and the minimum diffusion pump speed required to meet these performance figures. What is the system time constant?

6.2 The diffusion pump selected to meet the requirements of 6.1 has a speed at 10^{-2} mbar of 360 ls^{-1} and a critical backing pressure of 0.45 mbar. Determine the minimum speed required for the backing pump together with the minimum connecting pipe conductance. What pump-down time will elapse before the chamber reaches 10^{-2} mbar?

6.3 What pump-down time will elapse before the diffusion pump of 6.1 reaches constant speed if equality of throughput occurs at 10^{-1} mbar?

7

Construction accessories and materials

7.1 INTRODUCTION

Modern vacuum practice embraces basically two kinds of vacuum system: those based on glass and those based on stainless steel. The earliest vacuum systems were those constructed in glass, in later years, borosilicate glass (Pyrex). With the increasing complexity and number of attachments required with a vacuum system, it soon became impossible to build them from glass and recourse was made to stainless steel. At the present time glass systems are relatively less common, except in those cases where the system is designed to evacuate a glass envelope of some kind with a view to sealing it off. Under these circumstances it is convenient, although not vital, to use an all-glass system. In this chapter we shall review the materials and accessories associated with vacuum work, with an emphasis on stainless steel systems; glass-based systems will be covered in rather less depth. Construction technique will be interpreted very broadly to include important constructional elements, such as valves, cold traps and couplings. The introduction of motion into vacuum systems will be covered here as well.

7.2 STATIC PIPE COUPLINGS AND SEALS

For stainless steel vacuum systems the range of pipe couplings is quite extensive, with couplings being manufactured to cover the range of pressure from UHV to low vacuum. In recent years all these couplings have tended to meet a standard specification drawn up by the International Standards Organization (ISO). This standard is embodied by most manufacturers in their product range although there may well be differences of external detail.

The UHV fittings are based on the Varian 'Conflat'® design shown in Fig. 7.1. The sealing effect is a result of the geometry of the design which 'captures' the gasket material. The sealing gasket, an annular disc, is manufactured from OFHC copper and, when compressed between the conical sealing edges of the opposing stainless steel flanges, lateral cold flow occurs limited by the vertical flange wall. Material flow away from the seal area is severely limited and high interface pressures are developed which cause gasket material to fill surface imperfections and produce a highly reliable seal. Seals of this kind may be baked at high temperatures without loss of seal efficiency. The opposing flanges are brought together using high tensile strength, stainless steel bolts and nuts, lubricated with an anti-seize compound such as molybdenum disulphide. These seals and flanges are available in a wide range of sizes to fit tubing ranging from 19–356 mm in diameter. Flanges may be obtained rotatable, or double-sided if required, see Fig. 7.1 (b,c). Within the range of flanges designed to the 'Conflat'® pattern, there are two distinct classes; those designed for Imperial measure tubing and the ISO-Conflat® type, designed for larger diameter metric tubing.

In UHV work there are occasions when the Conflat® design is not suitable. This is particularly true for very large diameter apertures such as the mouths of vacuum pumps. In these cases recourse is made to wire seals using indium, gold (24 carat) or very high purity (99.99%) aluminium wire. In the case of indium, melting occurs at 156° C so bakeout above ~ 120° C cannot be contemplated. These wire seals may

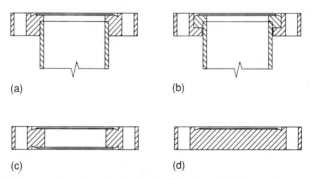

(a)   (b)

(c)   (d)

Fig. 7.1 Cross section through the Conflat® design of UHV coupling. (a) Non-rotatable. (b) Rotatable. (c) Double-sided. (d) Blank.

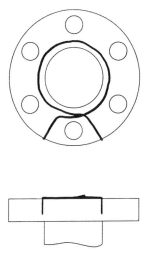

Fig. 7.2 Wire seal formed using a length of high purity gold or aluminium wire.

be obtained as preformed loops of the desired diameter, where the ends of the wire have been fused together end to end to form a smooth butt joint. This is not actually necessary if the wire material is really soft, e.g. Au or Al; the wire may simply be formed into a rough loop with the ends crossing over as in Fig. 7.2. When the plain polished surfaces of the opposing flanges are brought together on either side of the wire loop, plastic flow occurs at the crossing point and a perfect UHV compatible joint results. Seals of this kind may be used only once.

Usually, UHV compatible seals, i.e. seals which may be baked at 200° C and which will permit ultimate vacua of $< 1 \times 10^{-8}$ mbar, are confined to metal or wire gaskets. However, the UHV range can just be embraced using gaskets made from the fluoroelastomer Viton (Dupont), or the perfluoroelastomer Kalrez® (Dupont), if some limitations are accepted. Thus, Viton® seals should not be used at bakeout temperatures above 150° C, although Kalrez® seals may be baked at up to 250° C. These seals may be obtained preformed to fit standard Conflat® flanges, or as O-ring seals. Either way, they have the advantages of ease of use, reusability and capability of operating at pressures below 1×10^{-8} mbar; contrast copper gaskets which may only be used once.

For less exacting vacuum requirements, i.e. pressures $> 10^{-7}$ mbar, and this will normally include the backing pump line on any vacuum

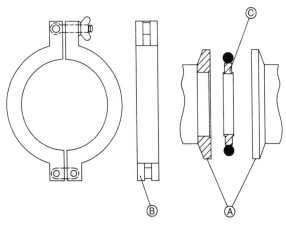

Fig. 7.3 Section and plan views of the KF (small flange) design. Here, A are small flanges; B is the hinged clamping ring, and C is the centering ring with O-ring gasket.

system, the most convenient coupling is the Klein Flange, KF (small flange), coupling introduced by Leybold of Germany and now an international standard. Figure 7.3 illustrates the essential features of this coupling, which relies on an O-ring seal made from fluoroelastomer or other synthetic rubber such as Perbunan for normal applications, although it is possible to fit a seal made from high purity aluminium if UHV conditions are required. Figure 7.4 shows a section through such

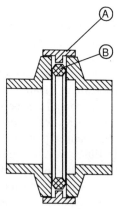

Fig. 7.4 Section through a KF seal designed for UHV work showing the centering ring A and the aluminium seal B.

a seal. The KF coupling has the advantage of being quick to assemble, reliable and bakeable to 200° C if necessary. Generally speaking, its use is confined to tubing sizes up to about 50 mm o.d. For larger size tubing, the KF fitting is abandoned in favour of yet another design due to the Leybold company of Germany, the ISO–K clamp flange.

The ISO–K clamp flange is depicted in Fig. 7.5. The essential differences between this flange and the KF are

1. size, the ISO–K being intended for larger diameter tubing;
2. the method of fixing, which now involves individual clamps spaced freely around the flange rim which engage in circular grooves in the flange.

While the KF flange may be secured merely by hand tightening a wing-nut, the ISO–K requires bolt tightening. The seal, though, is still effected by a rubber (fluoroelastomer) O-ring and yields vacua of around 10^{-8} mbar. If especially required, high purity, aluminium sealing discs may be fitted to yield ultimate vacua of < 10^{-8} mbar (UHV).

A very similar flange essentially identical to the ISO–K has fixing with conventional bolts and nuts through holes in the flange; this is the ISO–F.

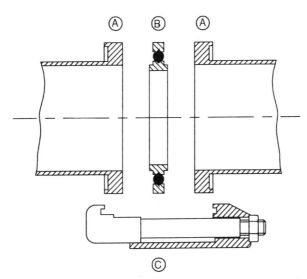

Fig. 7.5 Section through the ISO–K clamp flange; A are the ISO–K flanges; B is the centering ring and O-ring gasket with outer ring, and C is the clamp.

In vacuum work it is often necessary to build simple, quickly demountable fittings which do not match any of the international standards. For these purposes it is convenient to use O-rings with matching locating grooves of the appropriate profile. There are three main profiles; the regular groove, the trapezium groove and the muff coupling, see Fig. 7.6. The regular groove is suitable for horizontal surfaces only; the trapezium groove gives positive O-ring retention and may therefore be used for both horizontal and vertical faces; the muff coupling profile is primarily intended for sealing glass or other tubing into a vacuum system. O-rings are widely available in a large range of sizes and different varieties of rubber. For general vacuum work, neoprene or nitrile rubbers are adequate, although Viton® would be the generally preferred material since it has lower gas permeability than neoprene. Since rubber is incompressible, the cross sectional area of the groove should almost equal that of the chosen O-ring.

The muff, or compression sleeve, may be used as the basis of coupling systems which permit the coupling of modest diameter tubing. These may be metal-to-metal or glass-to-metal so that the coupling of a backing line pressure gauge, Pirani etc., may be readily and neatly achieved using the muff with a screwed coupling (Fig. 7.7). Alternatively the screwed coupling may be joined to a KF termination using an O-ring on a KF centering ring.

Although the muff to screwed coupling would be used to join glass to metal tubing, where the system is entirely glass it is simplest to design entirely without joints, except for the basic minimum of greased, taper joints in the backing line; the system then comprises entirely glass-blown joints. Glass UHV systems must be constructed this way. If a glass vacuum system requires to have demountable joints, these will normally be made using commercial ground taper joints lubricated and sealed with a low vapour grease such as Apiezon L, which has a

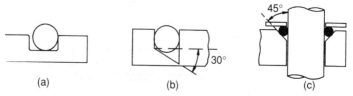

Fig. 7.6 Cross section through typical configurations for O-ring seals. (a) Regular grooves. (b) Trapezium groove. (c) Muff coupling.

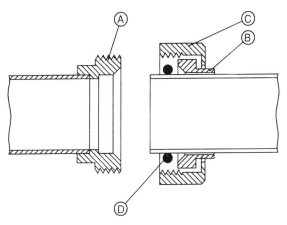

Fig. 7.7 Muff-to-screw coupling; A is the screw termination; B is the muff or compression sleeve; C is the coupling nut and D is the O-ring.

vapour pressure at room temperature of 8×10^{-11} mbar. In these cases ultimate vacua will usually be determined either by the vapour pressure of the grease or, more likely, the effective pumping speed of the system.

7.3 DYNAMIC VACUUM SEALS

A dynamic vacuum seal is required when it is necessary to introduce motion, either linear or rotational, into a vacuum system. Usually this is achieved using flexible stainless steel bellows as the sealing element although, where the pressure requirements are modest, say medium to high vacuum, it is possible to introduce both linear and rotational motion using rubber sealing rings (O-rings). There are other indirect methods of introducing motion into a vacuum which require no seals and are, in a way, pseudodynamic seals. These seals always require electrical connection or magnetic coupling in order to operate.

The simplest dynamic vacuum seal is usually based again on the O-ring. Figure 7.8 shows a cross section through assemblies designed to introduce either linear or rotational motion, or both, into a stainless steel vacuum system. The shaft which imparts the motion is sealed by O-rings located around the shaft section (Fig. 7.8(b)), or by shaped rubber washers (Wilson seals) (Fig. 7.8(a)).

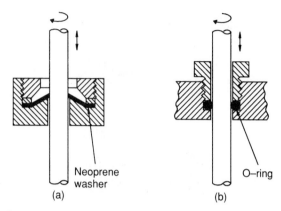

Fig. 7.8 Cross section through simple dynamic seals suitable for linear or rotating motion, or both. (a) The Wilson seal. (b) The O-ring seal.

For systems intended to operate under UHV conditions and therefore requiring to be baked at high temperatures, the O-ring seal is of no use. In all these situations the seal is effected by means of a thin-walled stainless steel bellows. Where linear motion is required, it is merely a question of selecting a bellows with sufficient flexibility to embrace the total linear motion required. Figure 7.9 shows a section through such a seal. Generally, seals of this sort have a calibrated thimble fitted to the

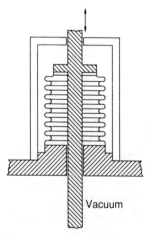

Fig. 7.9 Section through a seal designed to permit linear motion under UHV conditions. The mechanism which constrains shaft motion is omitted.

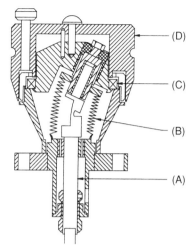

Fig. 7.10 Section through a dynamic seal suitable for introducing rotation under UHV conditions; A is the rotatable shaft; B is the stainless steel bellows seal; C is the external bearing and D is the calibrated thimble.

drive end which enables motion to be set with some reproducibility. Micrometer adjustment drives are available. A different approach is required to impart rotational motion to a vacuum shaft, however; the basic principle is shown in Fig. 7.10. The bellows is now bent askew to seal the cranked end of the vacuum shaft. The vacuum shaft is carried on ball bearings within the UHV side of the seal, while externally the bellows and the cranked end of the vacuum shaft are carried in a solid, but rotatable, sleeve which is drilled out to receive them. The end of the bellows is located accurately by a further bearing resting in the end of the rotatable block. Rotatable seals of this kind carry a calibrated thimble which allows a reproducible setting of the angular rotation to within a degree.

7.4 INDIRECT MOTION TECHNIQUES

While considering motion seals it is relevant to describe the indirect methods for introducing motion into a vacuum system. There are three principal methods. One, based on magnetic coupling, is normally applied to glass systems. Another, based on piezoelectric elements, is usually confined to metal systems. Both impart linear motion usually, although the magnetic coupling can be used to provide rotation. The third method provides the ultimate means of introducing rotational motion.

A most elegant indirect method of introducing motion into a vacuum system is that based on the Inchworm® motor developed by the Burleigh Instruments Corporation of the USA. Although originally conceived for the adjustment of optical elements, such as Fabry–Perot etalons, it was adapted in 1988 for use in vacuum systems, particularly those operating under UHV conditions. Figure 7.11 shows a section through such a motor. It comprises a shaft D, set around which are three piezoelectric sleeves A, B, and C, manufactured from barium zirconate titanate ceramic (PZT). Sleeves A and C are dimensionally identical and are poled so that the application of a voltage causes them to dilate or contract depending on the polarity of the voltage. The dimensions of these elements are set so that when contracted, they clamp rigidly onto shaft D. Sleeve B is poled so that application of a voltage causes a change in length, either an extension or a shortening. All three sleeves are cemented together, end to end.

The operation of the Inchworm® involves six steps repeated sequentially as follows. Sleeve A is contracted so that it clamps the shaft D, sleeve B is then extended and finally sleeve C is contracted to grip the shaft D. Sleeve A is now dilated followed by contraction of sleeve B. Finally, sleeve A is contracted. This completes a cycle of operations and the net effect is that sleeve A has moved to the right by an amount equal to the expansion of the central element B. This sequence of steps is repeated under automatic control, with a cycle repetition rate of 1 kHz, yielding motor element maximum traverse rates of 1 mms⁻¹ and a range, in UHV applications, of up to 200 mm. The mechanical resolution of the Inchworm® motor is 1 μm. This type of indirect motion controller may be operated either in open loop mode with no positional feedback, or in closed loop mode with a shaft encoder to establish shaft

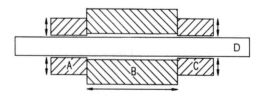

Fig. 7.11 Section through the Burleigh Inchworm® motor for producing linear motion. Piezo rings A and C may be clamped rigidly to bar D; piezo sleeve B may be extended or contracted. The arrows indicate the direction of movement.

position precisely. Because of its essentially ceramic nature it may be baked at 150° C and has a very low outgassing rate.

Magnetic coupling to produce motion in a vacuum is normally found in glass vacuum systems and usually amounts to moving a glass-encapsulated iron slug by means of an external permanent magnet. Most commonly, magnetic coupling is used for glass UHV valves. Figure 7.12 shows a schematic diagram of two types of magnetically operated valve, the Decker valve, Fig. 7.12(a) and a simple sliding mode valve, Fig. 7.12(b). These valves function effectively only at pressures such that molecular flow conditions prevail. An external permanent magnet is used to raise or lower the ground glass ball joint in (a) or to slide horizontally a carefully lapped glass disc in (b). Magnetically operated systems for introducing motion into stainless steel systems are also available, usually to produce linear motion.

Perhaps the most flexible, indirect, method for introducing motion into a vacuum system (linear or rotational) is that based on the stepper motor. Stepper motors are pulse driven motors which can rotate in fixed degree intervals, e.g. 1.8°. This makes them particularly suited to the precision movement of delicate specimens, even in UHV. They are available constructed entirely from UHV compatible materials (no lubricants) and capable of operating up to 150° C and bakeout at 200° C in background pressures as low as 10^{-11} mbar. Versions are available

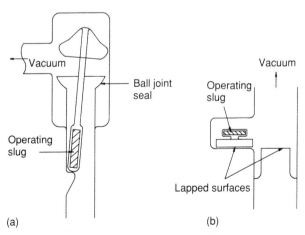

Fig. 7.12 Magnetically operated glass valves for glass UHV systems. (a) The Decker valve. (b) The sliding mode valve.

designed to give linear travel of up to 300 mm. They are perhaps the most flexible indirect means of introducing motion into a vacuum system although their use is confined to metal vacuum systems.

7.5 VACUUM VALVES

Vacuum valves divide into roughly five basic types:

1. the rubber diaphragm valve which is available in models to suit both metal and glass vacuum systems;
2. the tapered ground glass stopcock which finds its most ready use on glass systems;
3. the metal valve which is usually used on metal systems, but is also available to suit glass vacuum systems;

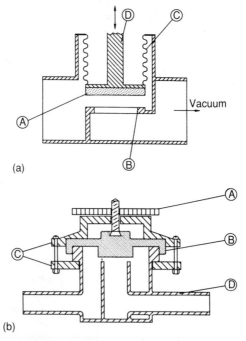

Fig. 7.13 Rubber diaphragm valves. (a) Metal body valves; A is the rubber sealing disc; B is the valve seat; C is the sealing bellows (stainless steel) and D is the actuating shaft. (b) Glass body valve; A is the actuating disc; B is the combined sealing disc and valve seal; C are the metal clamping rings and D is the tube to the vacuum system.

4. the ball valve usually confined to metal vacuum systems;
5. the gate valve, also confined to metal systems.

Most metal valves for vacuum work are available for manual or automatic operation simply by fitting an actuating solenoid. The mode of operation of the rubber diaphragm valve is shown in Fig. 7.13(a). A rubber diaphragm, A (generally Viton®), is mounted on a plate attached to a shaft, D, which can be raised or lowered above the valve seat, B, by means of a rotatable, threaded sleeve. The shaft is sealed off from the vacuum by means of a metal bellows, C. Valves of this type may be adapted for automatic operation simply by actuating the shaft, D, by a solenoid.

For glass vacuum systems a valve of essentially identical structure is available (Fig. 7.13(b)). In this case the mechanical part of the mechanism is sealed by the same diaphragm which is used to effect closure.

The metal valve, which was introduced by Alpert in 1951, is shown in essential detail in Fig. 7.14(a). The internal vacuum seal is made by pressing an OFHC copper gasket into a stainless steel knife-edge conical seat. The sealing pressure is great enough to deform the copper gasket on the conical seat. During bakeouts the loading mechanism counteracts thermal expansion and maintains the uniform pressure necessary for a good seal. Sealing pressure is supplied by means of a threaded shaft driving the copper gasket, the threaded shaft itself being isolated by means of a thin-walled bellows edge-welded to the valve body. In operation, the shaft will be tightened to a torque specified by the manufacturer. For glass UHV systems the metal valve operates in the same fashion, but the configuration of the bellows is now merely a thin plate with annular corrugations, rather than a tubular bellows (Fig. 7.14(b)). These valves have a conductance which may be varied continuously from ~ 0.3 ls^{-1} to 10^{-14} ls^{-1}. Valves with much larger open conductance are available. Where it is desirable to allow very close adjustment of flow rates, these valves are available with differential screw fittings which permit their operation as leak valves.

Ball valves are generally used in the pressure range from 10^{-5} mbar upwards. They are opened and closed by simple lever movement. A ball (stainless steel) is located within the body of the valve and bored out so that it opens or closes the pipeline according to the lever movement. In the open position, ball valves allow an entirely free flow; they have lubricated gaskets as valve seals. The essential features are depicted in Fig. 7.15. These valves are primarily intended for relatively high pressure applications on metal vacuum systems.

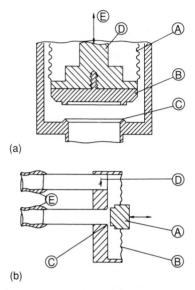

(a)

(b)

Fig. 7.14 Bakeable metal valves. (a) For stainless steel systems: showing A, the stainless steel bellows; B, the OFHC nose piece; C, the stainless steel knife edge and D, the operating shaft. (b) The metal valve for glass vacuum systems; A is the OFHC nose piece; B is the flat sealing diaphragm; C is the stainless steel valve seat; D is the outlet tube and E are the glass–metal seals.

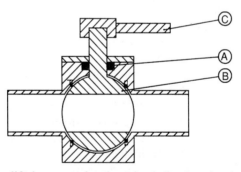

Fig. 7.15 Simplified cross section through a ball valve, showing A, the shaft seal; B, the rotating sphere and C, the operating handle.

In those situations where very high conductance valves are required, for example above the mouth of a diffusion or turbomolecular pump, the gate valve comes into its own, since in the open position there is no obstruction to gas flow and the diameter of the valve orifice may equal

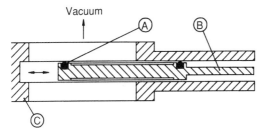

Fig. 7.16 Simplified cross section through a gate valve; A is the O-ring gate seal; B is the actuating shaft, and C is the seal housing.

that of the pump mouth. Figure 7.16 shows, in outline, the salient features of a typical gate valve. The actual gate is circular and is sealed by an O-ring set in one face. In operation the gate is slid into position from one side which houses the complete gate when no obstruction is required. The detailed design of the drive mechanism usually provides some means of avoiding sliding friction on the O-ring seal as it is moved from open to shut. Valves of this type may be baked to 150° C.

7.6 VALVELESS GAS ADMISSION

Although valves may well be thought of as providing the only means of controlling gas admission into vacuum systems, this is not so. For the gases hydrogen, helium and oxygen, there are additional methods available which are based on the fact that certain materials exhibit selective permeation. These materials are palladium or nickel (for hydrogen), quartz (for helium) and silver for oxygen. These materials are quite specific as to the gas and consequently result in the admission of gas of a high purity, often in excess of that obtainable using sealed bulbs of 'spectroscopically pure' gas. A further advantage is that the diffusion step in the permeation process is very sensitive to temperature, so that by controlling the temperature of the metal barrier, the amount of gas permeating the barrier may be adjusted very easily. Gas admission systems of this sort have no moving parts and are easily attached to bakeable UHV systems, or indeed, high vacuum systems.

In general, gas admission by selective permeation is performed by using an arrangement such as is depicted in Fig. 7.17. A thimble formed from the chosen metal is attached to the vacuum system via a

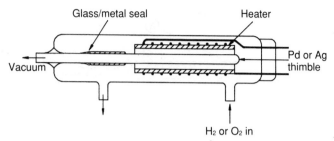

Fig. 7.17 Cross section through a metal thimble system for the controlled admission of very high purity gas, H_2 or O_2.

glass metal seal for a glass system, or by suitable brazing for a metal system. A nichrome heater coil is suspended around the thimble and the whole enclosed by the gas supply jacket. Nickel thimbles permeate hydrogen more slowly than those made of palladium, but are believed to be more reliable. Where it is desired to admit helium of very high purity (< 1 ppm), a quartz thimble ~ 150 mm long, 6 mm diameter and having a wall thickness of ~ 0.2 mm is suitable, but it needs to operate at temperatures around $750°$ C to provide, say, $0.1 \ \mu ls^{-1}$.

7.7 GLASS/CERAMIC–METAL SEALS

The fabrication of joints between metal and glass or metal and ceramic material is the basis of two types of important vacuum accessories, namely, optical inputs and electrical inputs. Vacuum systems which are primarily glass must have metal wires passing through the envelope in order to make electrical measurements within. Similarly, metal vacuum systems must have some way of insulating the wires passing into the system and transparent parts for observation purposes. The chemical and physical properties of glasses, ceramics and metals vary over a wide range so that a great deal of care must be exercised when joining these materials. Joints have to be, and remain, vacuum tight, yet still be capable of bakeout at high temperature if UHV is contemplated. Glass-to-metal seals have been used for many years as the basis of electron tubes and vacuum devices. In order to form a glass-to-metal seal, the glass must wet the metal. As a rule, glass does not wet clean metal surfaces but it does wet an oxide-covered surface, hence metals must be carefully oxidized before sealing to glass. The second problem involved in a glass-to-metal seal is matching the coefficients of expansion of the materials involved. This may be done in two different ways:

1. the matched seal, in which the thermal expansion coefficient of the metal is very similar to the glass to which it is attached;
2. the unmatched seal where the thermal expansion coefficients of the glass and metal are quite different, but the high stresses which would normally occur in such a joint are minimized by using either a very ductile metal or an intermediate graded seal terminating in a matching seal.

Glasses are now available which have thermal expansion coefficients approximately the same as the more common metals used in vacuum technology. In principle a suitable sealing glass can be found for any vacuum system metal.

One of the most common materials used in hard glass systems is Fernico or Kovar, an alloy of 54% iron, 29% nickel and 17% cobalt. It may be sealed directly to hard glass (Pyrex) and is available in many forms including electrical feedthroughs and optical windows. The structure of such a seal is shown in Fig. 7.18(b). Its drawbacks are that it is magnetic and a poor conductor of heat and electricity. A seal which is now less common, but which depends on the ductility of copper to overcome the mismatch in expansion coefficient between copper and glass, is the Housekeeper seal. Here, a feather edge is formed on the end of an OFHC copper tube or similar metal, and the glass sealed to this very thin edge. Typically, the copper tube will be tapered at ~1.25° to an edge of 0.05 mm, as shown in Fig 7.18(a).

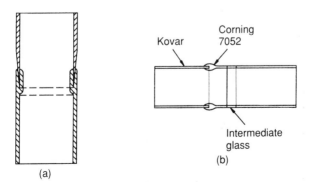

Fig. 7.18 Glass to metal seals. (a) The Housekeeper seal. (b) The graded seal.

Ceramic-to-metal seals are now more common than glass-to-metal seals; ceramics are stronger than glass, can withstand higher bakeout temperatures, have better dielectric properties and are less permeable to gases. There are four main categories of ceramic–metal seal:

1. diffusion seals;
2. sintered metal powder seals;
3. active alloy seals;
4. electroformed seals.

The diffusion seal is achieved by pressing a metallized ceramic ring to a metal partner at several hundred degrees centigrade.

In the sintered metal powder process, finely divided powders of tungsten, tantalum, molybendum, rhenium, iron or mixtures of these, are mixed with an organic binder and applied to the surface of the ceramic. The piece is sintered by firing in wet hydrogen at 1250–1500° C (depending on the ceramic). The resulting metallized ceramic is copper- or nickel-plated and then brazed to the metal part by conventional techniques using silver bearing braze alloy in a hydrogen atmosphere.

The active alloy process depends on the chemical reactivity of titanium and zirconium towards ceramics and glass. Both of these metals render ceramics wettable by reduction of the aluminium oxide. Titanium hydride suspended in nitrocellulose is coated onto the ceramic surfaces to be joined, with a thin washer of nickel or Cu–Ag eutectic interleaved between them. The whole assembly is fired in hydrogen, initially at 370–450° C, which reduces the titanium hydride to titanium which wets the ceramic surface, and finally to 850° C, which alloys the nickel or Ag–Cu eutectic to the titanium surfaces to form a vacuum-tight joint. The complete seal is thus made in one operation. Where it is desired to join just metal to ceramic, the combination would be ceramic/titanium hydride/nickel or Cu–Ag/metal. Suitable metals include tantalum, platinum, kovar, copper, and type 430 stainless steel (copper-plated). The titanium hydride process can also be applied to sapphire and quartz. Titanium renders quartz wettable around 600° C.

7.8 OPTICAL WINDOWS AND ELECTRICAL FEEDTHROUGHS

The main need for glass/metal or ceramic/metal seals arises from the wish to introduce either radiation (glass/metal), or electricity (ceramic/metal), into a vacuum system.

For optical systems, the viewport material must be chosen according to the transmission requirements of the experiment. It may well be that the viewport is required to do nothing more than provide an observation port, so that the state of affairs inside the vacuum chamber can be determined, in which case Pyrex glass will suffice as the window material. On the other hand, if the transmission properties of the window are important then use may be made of windows constructed from quartz, sapphire or magnesium fluoride, the last named having the widest transmission band. Regardless of the material actually chosen for the window, the manner of construction is much the same, the principal difference being the angle of the field of view available for a given port diameter. Figure 7.19 illustrates the design approach for two categories of window. Figure 7.19(a) shows a non-magnetic viewport achieved by using type 304 stainless steel reduced to a feather edge and sealed into a matching glass, Corning type 7056. This seal is, of course, a Housekeeper-type seal. No attempt is made here to give a wide angle of view. For a wide angle of view the seal shown in Fig. 7.19(b) would be used, where the metal in contact with the glass is Kovar and the seal is a matched expansion seal; this seal would however, be magnetic. The same basic principles apply if the port material is not a Pyrex type glass; so, for example, the window in Fig. 7.19(b) could be constructed of sapphire sealed into a Kovar ring using the titanium hydride techniques described in section 7.7

The introduction of electricity into a stainless steel vacuum chamber will always be carried out using a ceramic/metal seal designed along the lines shown in Fig. 7.19(c); if the chamber is made from glass, then

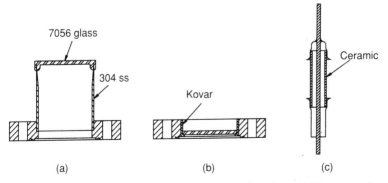

Fig. 7.19 The design approach for two categories of window. (a) Non-magnetic view part. (b) Zero length view part. (c) Electrical feedthrough.

a simple glass/metal pinch seal using tungsten or Kovar leadthroughs must be used.

Ceramics are a class of non-metallic inorganic materials which have been heat-treated with or without pressure to give them permanent shape and hardness. Ceramics include non-crystalline glass, glass-bonded crystalline aggregates such as porcelain and single phase compounds such as oxides, sulphides, nitrides, borides and carbides. The important physical properties of ceramics are their compression, tensile strength and thermal expansion coefficient. Alumina (Al_2O_3), which is the most commonly used ceramic for electrical feedthroughs, has a tensile strength roughly five times greater than that of glass; its compression strength is ten times greater than its tensile strength. This very high compression strength means that it is close to that of the metal it is being joined to; hence, a more rugged seal is possible than that which can be obtained with glass and metal.

There are nowadays a huge range of ceramic/metal seals designed to meet all conceivable requirements, ranging from passage of very large currents (600 A) or introduction of very high voltages (30 kV), down to screened, coaxial BNC feedthroughs which permit the measurement of very small currents. Specific feedthroughs are available for the connection of all types of thermocouples into a vacuum chamber.

7.9 BAFFLES, COLD TRAPS AND SORPTION TRAPS

An important element of any vacuum system is the baffle or cold trap. It is important to be clear about the uses of these two pieces of equipment. A trap (cold trap) is actually an entrapment pump for condensable vapours. A baffle is a device designed to condense pump fluid vapours and return them to the pump; it is therefore generally associated with diffusion pumps, which are the type of pump most able to produce oil vapour and oil vapour decomposition products. Although modern diffusion pump fluids such as DC705 or Santovac have vapour pressures in the region of 10^{-10} mbar at room temperature, some decomposition of the pump fluid does occur in the pump boiler and lighter fractions are generated. Many of these may be trapped by means of a water-cooled baffle situated above the pump first jet, Fig. 7.20(a). An even more effective trap is provided by liquid nitrogen cooling of such a baffle. A more generally useful arrangement is the liquid nitrogen cold trap situated immediately above the mouth of the diffusion pump, Fig. 7.20(b). In this position the cold trap not only holds

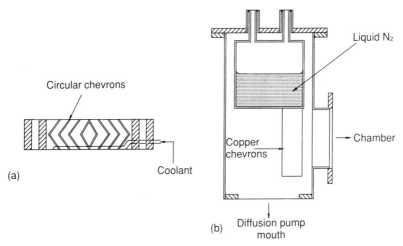

Fig. 7.20 (a) The water-cooled circular chevron baffle. (b) The liquid nitrogen cold trap.

the more volatile oil fractions arising from the pump, but also water vapour or other condensables arising from the vacuum chamber. The cold trap reservoir has a copper chevron attached at its base with a cross section similar to that shown in Fig. 7.20(a). This provides an optically opaque baffle between the chamber and the pump. Both water vapour and pump decomposition products are trapped by this sort of arrangement, yet the conductance can still be comparable to the speed of the diffusion pump, so that the overall pumping speed is still high.

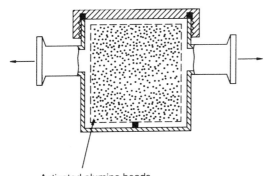

Activated alumina beads

Fig. 7.21 Cut away view of activated alumina sorption trap.

Finally, sorption traps are required to be interposed between rotary oil backing pumps and any main vacuum pump. It is always necessary to provide isolating valves so that the contents of the trap may be protected from contamination if the system is at atmospheric pressure for any reason. The structure of a typical sorption trap based on activated alumina is shown in Fig. 7.21. It comprises a vacuum tight container with a screw lid sealed by an O-ring seal; connection to the pumps is through diametrically-opposed tubes fitted with KF flanges. The interior of the container is a removable, perforated aluminium pot containing spheres of activated alumina. When the activity of the alumina is lost, it is generally best to replace the contents with a fresh, clean active charge. In desperation the existing charge can be reactivated by heating to around 250 °C for several hours.

7.10 METALS FOR VACUUM USE

There are two main classes of metal for use in vacuum system, namely, those intended to form the walls of the vacuum chamber, and those used to build substructures within the vacuum chamber. Which-ever category of use is envisaged, selected metals must meet certain basic criteria for vacuum use. These may be summarized as

1. They must have low vapour pressures at the highest operating temperature encountered.
2. They must have low outgassing rates.
3. They should be easily joined or sealed, either to each other or to ceramics.
4. They should not be porous or significantly permeable to atmospheric gases.
5. They should be strong.

Most metals have very low vapour pressures so that (1) above is readily met. There are, however, some common metals and alloys which should not be used in vacuum construction because their vapour pressures are high enough to interfere with normal bakeout procedures. Alloys which contain zinc, cadmium, sulphur, selenium or lead, for example, have vapour pressures which are too high for vacuum applications. As examples derived from the list above we might note that zinc is found in brass and readily sublimes from it on heating, cadmium is often used to plate steel screws and sulphur and selenium are constituents of free machining grades of type 303 stainless steel. These materials are excluded from modern vacuum practice. Also excluded are processing steps involving soft solder and brazing alloys requiring borax fluxes.

The most commonly used structural material is stainless steel, followed by aluminium. In many respects aluminium is preferable to stainless steel, but it is normally excluded from widespread vacuum use by the difficulty of making satisfactory welded joints and its inability to make a seal via a metal gasket. It has, however, the major advantages of low permeation rate, cheapness and ease of machining. In consequence, its use is restricted to those situations in which seals may be made using O-rings, or as internal fixtures for vacuum systems.

The preferred material of construction is stainless steel. It has a high yield strength and is easy to fabricate. The stainless steels used in vacuum work are selected from the AISI 300 series of austenitic steels; these are corrosion resistant, non-magnetic and easy to weld. The most commonly used alloy is type 304, although for very high quality construction type 316LN (a nitrogen-bearing alloy) would be used, which is more corrosion resistant, has greater weldability and higher yield strength. Particular care has to be exercised in both weld preparation and welding technique when stainless steel structures are being assembled. In order to avoid voids and inclusions, all welding must be executed on the inside of the chamber, a requirement that can impose difficulties, Fig. 7.22. All welding is done by the tungsten inert gas (TIG) process to avoid oxidation; the welding area is flooded with an

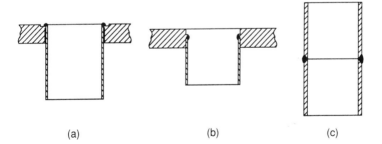

(a) (b) (c)

Fig. 7.22 Preferred methods of weld preparation designed to leave a crevice-free interior to the vacuum system. In each case, the black region represents metal fused during the welding process. (a) End groove preparation for thin-walled tubing. (b) Edge preparation for thick-walled tubing. (c) Butt welding of small diameter tube. For large diameter tube the weld would be made from the inside.

inert gas, usually argon. If the welding process is incorrectly carried out and the joint is overheated, carbide precipitation occurs, taking with it a substantial amount of the nearby chromium. Once the chromium content is reduced below 13% (normal stainless steel is 18% chromium and 8% nickel) it is no longer stainless steel and is subject to higher permeation rates for atmospheric hydrogen and is prone to micro-cracking. Over and above the preparation and welding of joints, it is vital that proper fabrication techniques are adopted so that inclusions are either avoided or rolled in directions that will not intersect a cutting or machining plane. Otherwise, for example, it is perfectly possible to have a pin-hole penetrating from one side of a flange to the other. Leaks of this sort are especially difficult to find since they frequently appear after the first bakeout, and, because they do not occur anywhere near a welded joint, they are seldom suspected.

During the conventional processing of stainless steels, the molten metal is allowed to solidify in moulds. This has the effect of cooling the metal from the outside in. In consequence, there is a variation in composition across the ingot, with the outer region often having greater density than the ingot core, which will contain impurities, cavities and general porosity. Non-metallic inclusions and low melting point alloy segregate at the core. If the ingot is then rolled, these inclusions, cavities, etc., form streamers along the core of the bar which lead to the problems outlined above.

The solution to these problems is to impose an additional refining process on the stainless steel known as electroflux refining or electroslag refining, in which the steel is remelted by means of the passage of a very heavy a.c. current through a consumable stainless steel electrode suspended above a layer of molten slag in a hearth. The droplets of molten steel fall through the slag which holds the majority of impurities in suspension or solution while the a.c. current provides a homogenizing, stirring effect. The resulting ingot is virtually pure and of uniform composition. The final step is to complete the shaping of the steel by forging alone which avoids streamers, followed by stress relieving.

Outside the range of stainless steel alloys used for major constructional elements, there are many metals which are suitable for structures within a vacuum system. The most commonly occurring ones are aluminium (dural) and copper, with smaller fitments involving tungsten, molybdenum, tantalum, gold, indium, platinum, rhenium, rhodium and nickel. Tungsten, molybdenum and rhenium are most commonly encountered as filaments in gauges, although both tungsten and molybdenum are pressed into use for those structures which must withstand high temperatures; for example, anode structures, heaters and heat shields. Gold has two classes of use, either as an O-ring metal gasket in UHV systems, or as a coating for electrode systems. It has a very low secondary electron emission coefficient and, in addition, it is largely inert; gas molecules have a very low sticking coefficient on gold.

Some of the important properties of materials which frequently find use in vacuum systems are gathered together in Appendix B.

8

Leak detection

8.1 INTRODUCTION

It is a fact of life that vacuum systems leak from time to time. Unfortunately, it is not always clear whether a real leak exists or not, since failure to reach the desired ultimate pressure may also be due to initial outgassing (virtual leak), gauge leakage current or other malfunction. If the equilibrium pressure attained is orders of magnitude away from the design figure it is almost certain that a real leak exists although, even then, it may be a failure (partial) of the pumping system. Thus, a diffusion pump operating with a fluid not specified for use in that pump can fail to boil the fluid properly, causing inefficient jet operation. This sort of problem may leave the pressure so high that many of the more precise leak-detecting techniques have to be abandoned.

In a new system it may take a considerable time (days) before the pressure reaches the design base pressure, so that the decision to search for leaks at all is as important as the method chosen for detecting them. In the extreme case, where the pressure in the vacuum chamber is absurdly high, it may well be possible to 'hear' the leak and anyway, under these circumstances, it will not be possible to operate the high vacuum pump. It is always necessary to adopt a clean and methodical routine in the assembly of a vacuum system, so that all individual components have been individually leak-checked before incorporation in the vacuum system and are free from surface contamination. This leak-checking will usually be done using a commercial leak detector of the kind described in Chapter 9, Section 9.2. The selection of a particular procedure for leak detection will usually be determined by the pressure ultimately attained in the operating system. Some procedures work best in a particular pressure range, whilst others may be completely excluded for operational reasons (Table 8.1). The most

Table 8.1 Principal leak detection methods

Technique	Leak range (mbar)	MDL* mbar (ls⁻¹)	Preferred test gas
Tesla coil	$2 \rightarrow 10^{-3}$	10^{-3}	
Pirani	< 10	10^{-3}	He, H_2
Ion gauge	$< 10^{-4}$	10^{-10}	He, H_2
Getter-ion	$< 10^{-4}$	10^{-11} (He)	He
RGA/PPA	$< 10^{-3}$	10^{-8} (H_2)	He, H_2

* MDL is the minimum detectable leak.

refined method of leak detection is that based on the residual gas analyser (RGA) or partial pressure analyser (PPA) such as were described in Chapter 5. For this reason it is good practice to have such an instrument incorporated into the vacuum chamber at the design stage. If the system pressure is not far away from the design figure, it is probably simplest to examine the RGA spectrum as a first step to see if it matches the typical residual gas spectrum for a leak-free vacuum system, or has unexpected peaks or enhanced peaks. Figure 8.1 shows a typical residual gas spectrum.

Leak detection is, almost without exception, performed on continuously pumped systems, i.e. dynamic, rather than static, leak detection. We can summarize the principal methods adopted in Fig. 8.2. This chart is not exhaustive as additional selective leak detection systems exist based on the palladium barrier gauge with hydrogen as test gas, or the oxygen detector with oxygen. These do not fall readily within the

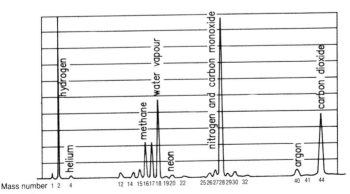

Fig. 8.1 Typical residual gas spectrum for an unbaked stainless steel vacuum system.

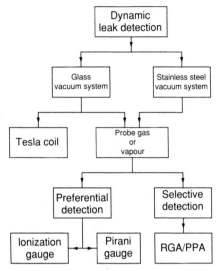

Fig. 8.2 Principal methods of dynamic leak detection.

scope of this book and will not be discussed further; they are also much less commonly used.

With the exception of the Tesla coil, all leak detection systems use a test or probe gas; in some instances vapours of acetone or ethanol are used. The throughput of the leak will change when it is exposed to the test gas by a factor depending on test gas viscosity or molecular weight. If viscous flow conditions exist then the change will be in the ratio viscosity of air/viscosity of test gas. When molecular flow exists, the ratio becomes $\sqrt{M_{air}}/\sqrt{M_{\text{test gas}}}$. Table 8.2 lists some of the relevant figures for common or potential test gases. Selection of a leak detection technique requires some idea of the basic technique sensitivity, or minimum detectable leak (MDL) flux. If this minima leak flux is Q_{min}, then we know that the minimum pressure of probe gas, p_{min}, which can be detected is given by

$$Q_{min} = p_{min} S_p',$$

Expressed another way, all leaks which produce a partial pressure greater than p_{min} may be detected. Now p_{min} will be as small as it possibly can be, i.e. maximum sensitivity is obtained if we reduce S_p' to zero; the leaking probe gas then simply 'accumulates' in the system. Consequently, this technique is known as the accumulation technique.

Table 8.2 Relative properties of probe gases for leak detection

Gas	Mol. wt	Mean mol. dia. (nm)	Ion gauge gas calibration factor*	Diffusion pump speed multiplication factor	Getter ion pump speed multiplication factor	Multiplication factor+ for changing air conductance at 20 °C to other gases	
						Molecular	Viscous
Air	~29		1.0	1.0		1.0	1.0
Hydrogen	2.016	0.275	3.1	1→2	2.7	3.8	2.07
Helium	4.003	0.218	7.7	1→1.5	0.1	2.7	0.93
Nitrogen	28.02	0.375	1.0	1.0	1.0	1.03	1.03
Oxygen	32.00	0.361	0.94	1.0	0.57	0.95	0.90
Argon	39.94	0.367	0.88	0.9	.01	0.85	0.83
Carbon dioxide	44.01	0.459	0.71	0.9	1.0	0.81	0.83
Mercury vapour	200.6	0.426	0.37			0.38	1.22

* Actual pressure = gauge indication × gas calibration factor.
+ Viscous flow through a leak occurs for throughputs > 10^{-5} mbar ls^{-1}; molecular flow through a leak occurs for throughputs < 10^{-8} mbar ls^{-1}.

The reader should perhaps be reminded of the basis of the 'leak' unit. Thus, a leak passing one standard cm^3 per second, that is 1 cm^3 at 1013 mbar, is equivalent to 1.01 mbar ls^{-1}, so that 10^{-11} mbar ls^{-1} is this quantity scaled down by 10^{11}. In practice leaks this small are rarely seen.

The maximum sensitivity described above can only be realised if the probe gas has had time to reach the steady state value. We can use the equation for the pump-down time to determine the time scale of this process for a volume V pumped by a pump of speed S'_p.

The pressure change p due to a sudden application of test gas to a leak is given by

$$p = p'_b + \frac{Q_{pt}}{S'_p}\left[1 - \exp\left(\frac{-S'_p}{V}\right)t\right] \qquad (8.1)$$

Here, p_b is the background pressure of the test gas and Q_{pt} is the throughput of the test gas. The original background pressure will be substantially restored in 5 τ where τ is the system time-constant. So if we reduce the system response time by increasing S'_p, the sensitivity is reduced and vice versa.

8.2 REVIEWING THE SYMPTOMS

Before leak detection techniques can be applied, it must be established that a leak actually exists. The straightforward way to do this is to plot pressure versus time with the system valved off from the high vacuum pumps. A leak will result in a linear pressure increase with time; outgassing will cause the pressure to rise non-linearly to an essentially stable value determined by the vapour pressure of the desorbing species. Figure 8.3 indicates schematically how these plots are likely to compare.

Sometimes this is not possible because the pressure obtained using the roughing or foreline pump by itself does not reach a value low enough to run the high vacuum pump. In this case there must be a gross leak, probably due to damaged or misfitting gaskets or insufficiently tightened flanges. This situation can also arise from seal failure in the backing pump, if it is a rotary oil pump, and it will become evident if the system can be isolated so that the Pirani gauge is reading only the pressure achieved by the backing pump. In other words, the rest of the system can be eliminated from consideration. A further problem, often

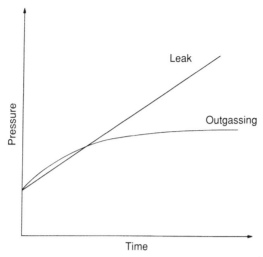

Fig. 8.3 Pressure/time characteristics for outgassing or leakage in a vacuum system.

associated with the backing pump line, is the condition of the activated alumina trap inserted to prevent backstreaming of pump oil products. If this trap is inadvertently exposed to air for any length of time, it will absorb a great deal of moisture which is liberated slowly during the early stages of pump-down. The vacuum obtained will then be poor, even though there is no real leak. Continuous pumping with gas ballasting will correct this problem, although a more certain cure is the replacement of the activated alumina charge with a freshly baked charge.

If the pressure can be reduced to a level such that the high vacuum pump may be operated, then the leak is of course smaller, or the outgassing less, or the vapour pressure of the contaminant less. In any event, as the operating pressure is reduced, other and more sensitive leak detection strategies may be operated.

It is uncommon, but not unknown, for leaks to exist through stainless steel flanges, otherwise almost all leaks in metal systems must be sought at joins between identical materials or the interfaces between dissimilar materials, e.g. glass–metal seals, ceramic–metal seals, etc. In glass systems, leaks may exist at points where glass-blowing has taken place despite the joint being perfect, or indeed invisible to the naked eye. Some, perhaps most, of the leak detection methods are common to both glass and metal systems. All methods are most easily employed if the vacuum is good enough to allow operation of the ion gauge, say, 10^{-4} mbar.

8.3 THE LEAK DETECTION ROUTINE

All successful leak detection depends on a totally systematic search avoiding, as far as possible, preconceived notions about the site of the leak. It is assumed that all components have been rigorously checked for leaks before assembly, using a standard leak checker based on a mass spectrometer tuned to the helium peak. The start-up of a new vacuum system will begin with the running of the backing pumps, generally rotary oil pumps, but sometimes sorption pumps, with the remainder of the system valved off. Usually the performance of this part of a vacuum system is faultless in the case of a system embodying new components; if, for example, the pump is elderly then problems may arise due to faulty operation of the pump. This will often be revealed by the changed 'note' of the pump, a hollow sound rather than the hard smack of a pump performing correctly. The pump will sound as if it is pumping air, and the pressure indicated by the Pirani or Convectron gauge will show that the pressure is inadequate to back, say, a diffusion pump. In the case of sorption pumps, the only 'fault' which can occur is that the zeolite charge has been allowed to absorb water and has virtually no pumping capacity.

If the backing or foreline system operates satisfactorily and produces a pressure sufficiently low for the start-up of the main bulk of the pumping system, then the valve connecting the backing pumps to the rest of the system may be opened. It is now a question of whether the backing pump can establish a vacuum in the remainder of the system. If it cannot, then there must be a major leak within the main section of the vacuum system, probably associated with an inadequately tightened flange or a damaged gasket, as described previously.

For a glass built vacuum system the procedure is much the same, i.e. systematic identification of those parts which are working correctly. Generally, though, it is more difficult to isolate sections of a glass system since the appropriate, convenient valves either do not exist, or where they do, would greatly reduce the pumping speed if used. (Chapter 7 provides details of glass valve types). On the other hand, the nature of construction of a glass system inevitably means that the opportunity for gross leaks (excluding cracks) is quite small, since glass blowing usually results only in pinholes, or minute leaks along glass–metal seals.

Once the backing pump system is operational the high vacuum pump may be started. In a new system the rate of pressure reduction to a level of around 10^{-6} mbar should be quite rapid, depending of course on the pumping speed and system volume, since these quantities control the system time-constant (section 6.5). We might note that the system pressure is reduced by half in $0.69V/S'_p$ seconds, or by a quarter in $1.4V/S'_p$ seconds and so on, where $V/S'_p = \tau$, the system time-constant. If V and S'_p are known for a given system, it is simple to check that this sort of behaviour is being produced.

A system which has been designed to reach an ultimate pressure which is of this order must be deemed satisfactory, although it may still be a prudent policy to monitor pressure versus time with the system isolated from the high vacuum pump if the design permits this. The pressure should rise slightly and then stick at a constant value if the system is leak free.

If a newly commissioned system fails to come close to its design pressure or, in the case of systems designed for UHV, shows little sign of the pressure tending to drop with time to 10^{-7} or 10^{-8} mbar, then the presence of a leak must be suspected. Again, this may be confirmed by isolating the high vacuum pump and monitoring the pressure with time.

In a glass system, leak detection may be carried out in several different ways, beginning with the use of the Tesla coil operated with just the backing pump. A Tesla coil produces a high voltage, high frequency discharge; 50 kV at 200 kHz is typical. The operational range of the Tesla coil is between 2 and 3 mbar pressure. The Tesla coil probe is moved over the glass surface, particularly in those areas in which glass blowing has been carried out. A glow discharge is obtained and, if the probe passes over a minute hole in the glass, a spark passing from the metal probe to the glass and hence to the discharge will be observed, the discharge being particularly located or pinned at the hole in the glass. The position of the hole may then be marked and further glass blowing carried out to effect a repair. Some caution must be exercised in any areas where the glass is known to be thin, since the Tesla coil is capable of piercing such glass. A particular feature of glass vacuum systems is the stopcock, as described in Chapter 7. Leaks occur on these, generally due to tracking in the grease layer, but the Tesla coil may again be used to establish leaks of this sort. Where glass-to-metal seals exist, the Tesla coil cannot be used as the discharge is automatically located on the metal component and no useful information is obtained. An alternative approach is to set up a glow discharge in an adjacent part of the system using the Tesla coil and then blow a fine stream of carbon dioxide over the test region, where the glow discharge will change colour to blue-green as the carbon dioxide replaces the gases in the vacuum system. Perhaps a more convenient alternative to carbon dioxide is ethyl alcohol, again sprayed on to the suspect area.

In all systems it is possible, in desperation, to make a guess at the location of the leak and make a temporary seal. If the guess is correct then the pressure will fall. Temporary seals may be executed using Q compound, a plasticine-like material, or particularly in the case of glass systems, simply by spraying with distilled water which will freeze in the hole under the continuous pumping from within the system. This seal or plug is ultimately pumped away and the pressure rises again. Water tends to seal holes that are too large to be sealed by acetone or ethanol.

8.4 LEAK DETECTION USING A TEST GAS

In dynamic leak detection we elect to probe the surface of the vacuum system with a probe gas and detect the presence of this probe gas either preferentially or selectively. A very important example of a detector of the first kind is the ionization gauge. Now, the probe or test gas should have a viscosity which is noticeably different from air so that its entry to the vacuum system is markedly different also. Secondly, it should be pumped by the high vacuum pump with a speed different from that of air and finally, the response of the detector should be different. Of course, some care must be taken that these factors do not balance out.

The ionization gauge has a sensitivity which is dependent on the nature of the background gas, and so for that matter does the Pirani gauge for the higher pressure regime. This fact is used together with a probe gas which is directed in the form of a fine jet over the walls of the system. A small cylinder of the chosen gas (helium is safe and effective) is connected via rubber tubing to a fine jet of either glass or metal, and a jet of gas passed over the system at a rate of about 1 cm in 10 seconds. The response of the ion gauge depends on the gas chosen as the probe and the type of high vacuum pump in use. With helium or hydrogen as probe gas and a diffusion pump based system, the pressure will actually drop when the probe gas locates the hole since the pumping speed for these gases is much higher than for air, and this outweighs the other factors operating.

The physics of the situation is readily understood by reference to Fig. 8.4. Consider first the passage of air through the leak with a throughput Q_A, yielding a pressure p_A when the pumping speed for air is S_A. Now, if the air is replaced by helium the throughput becomes Q_{He}, the pressure (actual) p_{He} and the pumping speed S_{He}. In the molecular flow regime, which is assumed for this level of leak, $Q_{He} = 2.7\,Q_A$ and $S_{He} = 1.5S_A$ from Table 8.2, so that if we calculate the ratio of the pressures, p_{He}/p_A, we get

$$\frac{p_{He}}{p_A} = \frac{Q_{He}}{S_{He}} \frac{S_A}{Q_A} = \frac{2.7}{1.5} = 1.8 \qquad (8.2)$$

On the face of it the pressure reading appears to increase by a factor of 1.8, but we have neglected to incorporate the gauge calibration factor for helium of 7.7, so that p_{He} (indicated) is equal to $p_A/7.7$. If this is taken into account then what we see is that the indicated pressure ratio will be 1.8/7.7 = 0.23, a pressure drop. A similar, though somewhat less marked, effect is obtained for hydrogen. Although this calculation works reasonably well for a diffusion pumped system it is less certain for a getter-ion pumped system where the pump speed is highly selective, being determined by chemical, rather than physical factors, so that the pumping speed of helium is less than a tenth of that for hydrogen, for example. If a different gas is chosen, e.g. butane, then the pumping

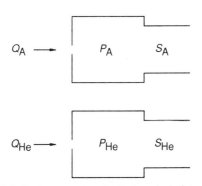

Fig. 8.4 Basic parameters for test gas leak detection.

speed is less and the gauge sensitivity higher so that the pressure indication rises. In the range from 10 mbar down to 10^{-3} mbar, when pumping on just an oil-sealed rotary pump, the Pirani gauge used in conjunction with hydrogen as the probe gas, is a sensitive combination. Alternatively, if the rotary pump is backing a diffusion pump, then the Pirani gauge will pick up leaks in the pressure range 10^{-3}–10^{-7} mbar measured above the diffusion pump.

Where a system is pumped by a getter-ion pump, the current passing through the gaseous discharge in the pump is a linear function of the pressure in the pump so that no additional vacuum gauge is required. Generally, getter-ion pumps are supplied with a power unit already calibrated in pressure terms so that it is merely a question of observing the pressure reading as the probe gas is passed over the suspected leak. If the probe gas is helium or nitrogen the discharge current (pressure reading) increases, while if it is oxygen or carbon dioxide it decreases; hydrogen is not a suitable choice here.

Where a system is pumped using a turbomolecular pump, which has a low pumping speed for hydrogen, the response of the measuring system will be the opposite of that found for a diffusion pumped system, i.e. the pressure will apparently rise when the hydrogen probe jet encounters the hole.

Although the techniques described above rely upon the use of probe gases, it is often possible to obtain the same information using nothing more complicated than a wash-bottle filled with acetone. The acetone is sprayed onto the area suspected of leaking and, if a leak actually is present, the ionization gauge, or for that matter a Pirani gauge, further back in the system will register a transient pressure rise. This approach, although cheap and simple, is less controllable than the gas jet approach and has the special disadvantage that the acetone may freeze in the leak. It may be used as a last resort.

The basic ideas described above may be employed in a different way which involves delivering helium to whole sections of a vacuum chamber. This will tend to be done when the probe gas approach has failed to locate the leak. Because helium is less dense than air, it is possible to arrange to 'bag' the chamber, or sections of it, with large polythene bags into which the helium is introduced. Air is gradually displaced from the bags by the helium so that if a leak exists in that region, its presence will eventually be revealed. It is then a question of restricting, in successive stages, the volume of the system which is enclosed by the polythene bag.

8.5 LEAK DETECTION USING PARTIAL PRESSURE ANALYSIS

A system which is intended to attain UHV may well appear to perform satisfactorily in the initial stages of pump-down, and even reach 10^{-8} mbar without bakeout. After bakeout though, it may be found that the vacuum simply does not improve and that there is a minute leak. The smaller the leak, of course, the more difficult it becomes to find it. For this reason it is always advantageous, and indeed good design practice, to fit a simple mass spectrometer to the vacuum system as was pointed out previously. Nowadays, the choice would be one of the many compact quadrupole designs. The advantage that this brings is easily seen if one contrasts the minimum detectable leak achievable with each system, thus the Bayard–Alpert ion gauge can detect a leak as small as 10^{-7} mbar ls^{-1} while the RGA can manage 10^{-11} mbar ls^{-1}, and under special conditions, 10^{-13} mbar ls^{-1}. Now the RGA will be tuned to detect helium and it has a threshold sensitivity to this gas. In a typical operating system, a residual background pressure of this gas will be present either because it back-diffuses, is lost from the cold trap, or is regurgitated by the pump. This residual background may well be greater than the ultimate sensitivity of the detector and will increase the minimum detectable probe gas pressure. When the leak is probed with helium, the partial helium pressure due to the leak builds up to an equilibrium value, p_{He}, given by

$$p_{He} = \frac{Q_{He}}{S_{He}} = \frac{2.7}{1.5} \frac{Q_A}{S_A} \simeq 2 \frac{Q_A}{S_A}$$

The minimum detectable leak Q_A (min) is then given by

$$Q_A \text{ (min)} = 0.55 S_A \, p_{He} \text{ (min)} \tag{8.3}$$

Hence the minimum detectable leak (MDL), Q_A (min), is proportional to pumping speed and the minimum partial pressure of helium that the instrument can detect. This result is true for pressures below about 5×10^{-6} mbar which sets a limit on the amount by which the pumping speed can be reduced.

PROBLEMS – CHAPTER 8

8.1 A vacuum chamber is pumped with a diffusion pump having a speed at the vacuum chamber of 1000 ls^{-1}; the lowest pressure which can be reached is 10^{-6} mbar. The system is probed for leaks using a helium jet. What percentage pressure change will be recorded when a leak is located?

9

Vacuum systems in practice

9.1 INTRODUCTION

The preceding chapters have surveyed the basis of vacuum physics and techniques ending with a brief survey of leak detection methods. In practice, of course, this is the real end point, since before a new vacuum system can be brought into use, leak detection and checking must be carried out. However, it seems appropriate to finish with a chapter which surveys vacuum systems in practice since this makes it possible to show how the ideas presented in earlier chapters are brought together in operating vacuum systems and, at the same time, emphasizes the full scope and real complexity of the modern vacuum system, which is not explored to any extent in the previous chapters where we have considered somewhat austere vacuum systems. Any choice of system must, of course, be entirely arbitrary, so three systems have been selected which exhibit special features spanning the range from the mundane to the exotic, the practical workhorse to the research tool. In the first category it seems appropriate to examine the workings of a commercial leak test system. This is useful for at least two reasons. Firstly, it provides continuity with the material discussed in Chapter 8 and, in effect, expands our view of how leak detection may be carried out. Secondly, it provides an example of a vacuum system which, although involving all the physical concepts described in earlier chapters, namely pressure gauges, magnetic deflection mass spectrometers, vacuum pumps and electromagnetically operated vacuum valves, is yet entirely portable (may be lifted by one, or more comfortably, two people). In other words, it may be carried to the leak. It is also an example of a fully automatic system, with microprocessor control.

The second choice of vacuum system is more rooted in technology and demonstrates the practical bell-jar system used for the routine deposition of materials ranging from metals to insulators, from aluminium to diamond, and is the sort of vacuum system widely used commercially for specialist coatings. The means employed for generating the coating may range from electron beam impact through simple sputtering to plasma based systems. Much of the physics involved has, as it happens, already been introduced in the description of the mode of operation of ion pumps. The third choice is taken from the realms of surface science research and is of particular interest, since it is equipment used to explore the physical details of the interaction between gas atoms/molecules and solid surfaces. Chapter 3 has discussed, to some extent, the important role which surfaces play in a vacuum system, and how their essentially 'sticky' nature with respect to gas interactions can dominate vacuum system behaviour or, alternatively, can be used as the basis of pumping systems. The system to be described provides an example of the most sensitive technique available for the examination of surfaces and the interaction of gases with surfaces. More interestingly though, it is an example of a system, which, in effect, operates with an enormous, but designed leak; in consequence, the range of pressures encountered in this research instrument spans nearly eighteen decades from 200 bar down to 10^{-12} mbar, the widest range of which the author is aware in any vacuum system. It has another characteristic, namely, the use of not just one vacuum pump but many.

9.2 THE PORTABLE LEAK DETECTOR

The equipment selected as an example of a portable leak detector is the Veeco MS–20 system manufactured by Veeco Instruments Inc. of the USA. This portable system automatically measures time and gross leaks in the range from 6×10^{-10} standard $cm^3 s^{-1}$ (a standard cm^3 is one cm^3 at STP or 1.013 mbar litre) down to 10 standard $cm^3 s^{-1}$. In order to obtain easy portability and use, the system is designed so that no cooling water supply is required for the diffusion pump and no liquid nitrogen is required for cold traps. A single direct-drive mechanical pump provides not only the backing vacuum for the diffusion pump but also the roughing vacuum for the test manifold. This is, of course, quite conventional. Rather less conventional is the use of an air-cooled rather than a water-cooled diffusion pump. It is the use of a fan-cooled diffusion pump which provides the freedom from water supply lines.

The automatic operation of this vacuum system is ensured by the combination of full microprocessor control with eight solenoid operated vacuum valves of the general pattern shown in Fig. 7.13(a). These valves may also be controlled manually. The actual leak detection is achieved using helium as the test gas and a high resolution, dual 90° sector magnetic analyser mass spectrometer.

The layout of the vacuum system is shown schematically in Fig. 9.1, where it can be seen that the structure may be broken down into seven basic elements or groups of components, namely, the mass spectrometer tube, MS, the solenoid valve test manifold assembly, the mechanical pump, the diffusion pump, vacuum gauges, and finally the high vacuum and roughing pipelines. Associated with each group of operating components such as a vacuum gauges, solenoid valves or the mass spectrometer, there is a corresponding electronic control board which interfaces with the microprocessor, as shown in Fig. 9.2. Beginning with the direct-drive rotary pump, this may be isolated from the

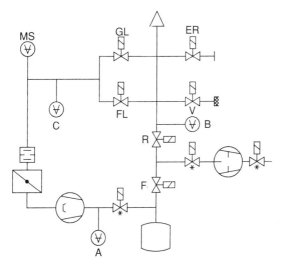

Fig. 9.1 Schematic diagram of a portable leak detection system based on the Veeco MS–20. Solenoid operated valves control gas flow under computer control. They comprise a gross leak valve, GL; a fine leak valve, FL; an external roughing line connection valve, ER; an air vent valve (filtered), V; a roughing line valve, R and a foreline valve F. Valves labelled with asterisks afford protection against spillage of oil during transport. The vacuum gauges comprise Pirani gauges A and B; Penning gauge C. The mass spectrometer is denoted by MS.

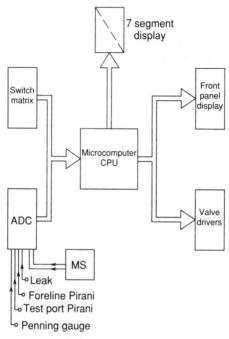

Fig. 9.2 Organization of microcomputer control for an automatic portable leak detector.

rest of the vacuum system and also from the atmosphere by means of two valves which may be thought of as 'shipping' valves in the sense that they serve to protect the rotary pump and the rest of the vacuum system from oil spillage during transport. A solenoid valve with the same function is fitted on the outlet of the diffusion pump. In addition to these valves the backing pump inlet line or foreline contains two further solenoid operated valves which serve to isolate either the diffusion pump or the inlet manifold from the rotary oil pump; these constitute the foreline valve F and roughing valve R respectively. Within the foreline is a ballast tank which may be isolated from the rotary pump. This is a useful device for any system in which it is necessary to use the backing pump to pump down an ancillary pipeline from atmospheric pressure, in this case the test manifold. The vacuum in the ballast tank provides the backing vacuum for the diffusion pump during the time required for roughing out the test manifold. Obviously, rough pumping of the test manifold cannot be maintained indefinitely.

Pressure measurement in the vacuum system is performed by two Pirani gauges and a Penning gauge. The Pirani gauges, A and B, are fitted to the foreline and roughing line respectively while the Penning gauge, C, provides the means of measuring the high vacuum attained in the mass spectrometer region. The mass spectrometer in this instance is comprised of two 90° sector magnetic mass separators tuned to the helium peak.

The operation of the system as a leak detector may be carried out in any of three different ways. The first method is 'sniffing', in which the chamber or device under test is filled with helium or a mixture of helium and nitrogen at a pressure greater than atmospheric, when the helium will escape from leaks and may be detected using a probe connected to the leak detector under fine test conditions (Fig. 9.3(a)). The probe is moved slowly around the vessel being tested; minute leaks can be located in this manner. The second technique, which is in effect the inverse of this approach, is to connect the chamber, if it is small, to

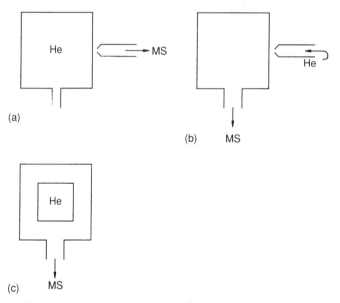

Fig. 9.3 The principal operating modes for leak detection using an automated leak detector. (a) System filled with He, mass spectrometer MS 'sniffs' for leaks. (b) Mass spectrometer MS connected to vacuum equipment, He jet 'sprayed' over external surfaces. (c) Equipment impregnated with He under pressure ('bombing'). Any trapped He escapes and may be detected.

the test manifold of the leak tester and use a fine jet of helium directed systematically over the joints of the test system. This approach, although convenient, is less precise in its ability to locate leaks since helium may readily diffuse through the atmosphere and give a test response when the probe is merely adjacent to a leak; this technique may be thought of as 'spraying' and is illustrated schematically in Fig. 9.3(b).

The third approach to leak checking using a leak detector is to take the chamber or device under test and fill it with helium and place it in a chamber connected to the test port of the leak detector. This chamber should have a volume not greatly in excess of that of the test item so that the pump down time is not excessive. Of course, it is often the case that it is not possible to introduce helium in this way so the alternative is to subject the item (small) to a high pressure helium atmosphere before inserting it in the test chamber. If leaks are present in the test item, then the high pressure will force helium into the test item and this helium will subsequently escape and be detected. This approach is known as 'bombing' and is suitable for devices or items which are normally hermetically sealed (Fig. 9.3(c)). Generally speaking the choice of technique is dictated by the size of the test item, thus a large vacuum chamber may be tested for leaks using the sniffing technique, Fig. 9.3(a), while it is only smaller items which may be checked using the 'spraying' technique, Fig. 9.3(b), or the 'bombing' technique (Fig. 9.3(c)).

Reference to Fig 9.1 again permits a more detailed description of the mode of operation of the leak detector. The solenoid valves labelled with an asterisk are the 'shipping' valves described earlier. These valves isolate the inlet and exhaust ports of the rotary pump and the backing line of the diffusion pump. These valves operate (open) automatically when a.c. power is applied to the unit. Another valve which may be regarded as a 'shipping' valve, in the sense that it protects and isolates the input to the diffusion pump, is the butterfly valve at the inlet to the diffusion pump. This valve is not under automatic control and may be operated manually to facilitate work such as cleaning and maintenance of the mass spectrometer or Penning gauge whilst the diffusion pump is still operating.

When the equipment is first switched on, the diffusion pump cooling fan starts. This is followed by switching on the roughing or foreline pump and the control electronics. The unit then tests that all elements are operating correctly and if this is so, indicates audibly. The diffusion pump may be switched on when the foreline pressure is less than 5×10^{-2} mbar. The untrapped three-stage diffusion pump is intended to produce a high vacuum of around 1×10^{-6} mbar. This is the operating pressure, which will be achieved in around ten minutes. During operation, the pressures within the various sectors of the system are monitored continuously by the vacuum gauges which in turn operate the solenoid valves via the microprocessor. Thus the foreline Pirani, A, monitors the pressure at the outlet of the diffusion pump and will stop the leak-checking routine and reopen the foreline valve if the foreline pressure exceeds 3×10^{-1} mbar. If the pressure rises as high as 4×10^{-1} mbar, it will cause the diffusion pump to be turned off. The manifold Pirani, B, is used to monitor pressure from 1 atmosphere to 1×10^{-3} mbar and controls the changeover from gross test to fine test on the leak checking procedure. This will be described in more detail later. Finally, the Penning gauge, C, monitors the pressure at the diffusion pump inlet (in the range 1×10^{-3} mbar to 1×10^{-6} mbar) and also in the mass spectrometer tube. If the pressure in this region rises above 9×10^{-4} mbar, the test valves are closed automatically and the filament of the mass spectrometer turned off.

The test manifold is fitted with four solenoid operated valves. Two of these valves are concerned with the start-up and completion of operations. The vent valve V is fitted with a filter and permits the inlet of air to the system at the completion of testing. The external roughing valve, ER, will normally remain closed unless it is required to use an external mechanical roughing pump. Central to the leak test operation, however, are the gross and fine test valves which connect the mass spectrometer tube to the system being checked. These valves are operated according to a pressure schedule so that the gross leak valve opens when the manifold pressure is less than 5 mbar and the crossover to the fine leak valve occurs at a pressure of 5×10^{-2} mbar. In the gross leak mode, one part in 100 000 of the leak is sampled into the mass spectrometer tube and the spectrometer signal multiplied electronically by 10^5. The crossover pressures may be set by the operator, but the system contains built-in protection against the selection of potentially damaging crossover points.

Although the operation described above is automatic and controlled by the microprocessor, manual operation of all the valves is possible. However, in this mode the microprocessor checks the validity of all the valve settings and permits only those which will fit in with the normal leak detection operation.

9.3 VACUUM-BASED COATING SYSTEMS

The discussion of vacuum system structures within the preceeding chapters has been centred on somewhat basic, totally enclosed vacuum systems. A class of vacuum system which is very important industrially and technologically, is that based on the removable vacuum chamber akin to a bell-jar. Systems of this type may nevertheless have a large capacity working chamber and the option for partial bakeout of the removable chamber together with the type of automatic control facilities described for the leak checking system of section 9.2. The usual role for such systems is the production of specialist coatings which may themselves range from metal films, through insulating films, to wear-resistance films; the range is enormous. It will include simple metal films such as Al, Ni, W or Pt through superconductors perhaps like NbN, passivating and insulating films, for example SiO_2, Al_2O_3 or Si_3N_4, wear-resisting films like TiN, TiC, WC or even diamond. All of these materials require a vacuum environment for their production; the difference will lie in the choice of film-generating technique. Technologically they are very important. An example of a bell-jar type vacuum system which permits coating operations is the Edwards E610A system which is designed for production coating in the electronics and optical industries, although it is sufficiently flexible for other applications. The basis of the design is a versatile, vertical stainless steel bell-jar unit of substantial size, 720 mm high by 610 mm diameter. This bell-jar has a highly polished interior, viewing ports and optional water-cooling. It is sited on a stainless steel baseplate and sealed to it by means of a Viton ring seal. The baseplate carries the necessary feedthroughs for electrical equipment. The bell-jar can be raised or lowered by means of a hydraulic hoist.

Pumping requirements of the vacuum chamber may be met using a variety of vacuum pump combinations ranging from the Edwards Diffstak® type diffusion pump through to turbomolecular pumps or cryopumps, depending on the particular quality of vacuum required; the cryopump of course, offers the cleanest deposition environment. The residual gas traces of the different pumping systems has already been examined in Chapter 4. A standard rotary backing pump is, of course, required in addition.

The vacuum system is provided with a fully automatic pump-down sequence operated by a single start button. Alternatively, microprocessor control may be used where a higher level of automation is required. In this respect the basis of control would be similar to that outlined for the automatic leak detection system of section 9.2. A guide to performance may be obtained from the pumping speed of the pumps usually fitted which, in the case of the Diffstak®, is 1700 ls⁻¹, with similar speeds for the alternative pumping systems. The Diffstak®, it should be pointed out, is a patented design due to the Edwards High Vacuum Co. which provides a high performance diffusion pump which does not require extra traps and baffles or indeed need liquid nitrogen for really clean pumping. It represents an improvement in subtle ways over the basic diffusion pump structure, it incorporates a water-cooled baffle to prevent backstreaming and runs at a lower operating temperature than usual, which minimizes the production of high vapour pressure, light oil fractions.

The central feature of any coating system is, of course, the means used to provide the coating. The E610A system is able to accept a variety of coating sources, ranging from electron beam heating to r.f. excited plasma sputtering-based systems. The physics of these coating systems has, to a large extent, been described in a different context, that of the ion pump. In the case of electron beam heating, this has emerged in the discussion of ion gauge outgassing. The particular advantage of electron beam heating (e-beam) is that it enables evaporation to occur from a source without contamination from the containing crucible or heater. In some cases, silicon for example, the molten charge in the crucible will readily attack and alloy with the crucible itself. Figure 9.4(a) shows the basis of the e-beam approach. A thermionic filament supplies the current to the beam and electrons are accelerated by an electric field to around 500–1000 eV and strike the surface of the metal charge to be evaporated, for example, aluminium. To prevent impurities from the filament reaching the aluminium charge, and more importantly, to remove the electron source from the line of evaporation of the aluminium a magnetic field from a built-in permanent magnet bends the e-beam. The energy of the impacting electrons is transferred to the aluminium charge which is maintained in a water-cooled hearth. Evaporation of aluminium takes place from a molten hot spot occurring at the point of electron impact. This type of approach is particularly useful for difficult materials such as silicon and means that very clean silicon deposits can be made. If necessary, co-evaporation of two materials can be made simultaneously using a double hearth facility.

Perhaps the most useful general means of evaporation is that based on sputtering, as described in Chapter 3. Sputtering may be carried out using energetic charged particles, Ar ions, or in some ways more usefully, by fast neutral atoms formed by charge exchange. Most commercial sputtering systems offered for use in a vacuum bell-jar are

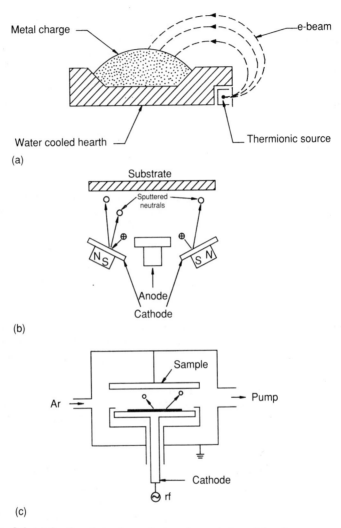

Metal charge

e-beam

Water cooled hearth

Thermionic source

(a)

Substrate

Sputtered neutrals

N S   S N

Anode

Cathode

(b)

Sample

Ar

Pump

Cathode

rf

(c)

Fig. 9.4 (a) Section through an electron beam (e-beam) heating system. (b) Conical magnetron structure for sputtering with argon ions (c) Radio-frequency plasma etching system using argon.

based on the magnetron principle. Figure 9.4(b) shows a section through a conical magnetron which has a concentric anode and circular symmetry. Electrons originating at the cathode are confined by the fields of the permanent magnets and are collected by the anode. The argon ions formed by electron impact do the actual sputtering. A large fraction of the sputtered cathode material, which is composed of neutral atoms, is ejected forward and deposited on the substrate. The magnetron operates at voltages of around 100 V, an order of magnitude less than that required for e-beam heating.

A very widely used sputtering system is that based on formation of a plasma excited by r.f. The electron concentration in the plasma is relatively low, of the order of 10^9–10^{12} cm^{-3}. In the r.f. driven system, Fig. 9.4(c), the powered electrode becomes negatively biased during operation and thus constitutes a cathode to which the positive ions in the plasma are attracted. The potential drop they experience is of the order of 500 V, the cathode fall, previously described in Chapter 3. The incident ions are accelerated by this potential onto the target material and sputter target material onto the sample to be coated mounted immediately above.

It was pointed out above that sputtering using fast neutral atoms is, in some ways, much more useful since the sputtering of insulating materials can be performed without consequent sample charging. There is a wide range of very important insulating materials which may be deposited in this way from sputtering sources having a large fast neutral atom content. Perhaps the most important of these materials is diamond. Although the coatings eventually layed down are not necessarily true diamond in structure, they are certainly diamond-like and may be made to adhere strongly to a range of materials as diverse as Ge, Si, Al, Ti, glass and polyethylene. A convenient way of achieving this objective is by use of a source intended to supply primarily fast atoms.

A particularly interesting, yet simple example of such a sputtering source is that based on saddle-field confinement. Sources of this type are compact and reliable; they are capable of producing ions or fast neutrals in a naturally collimated beam. Indeed it is this type of source, which, when fed with butane, will deposit a diamond-like surface coating.

The saddle-field ion/fast neutral source is an example of a device in which electrons are confined to a limited volume solely by means of an electrostatic field, in apparent contravention of Earnshaw's theorem. Physically, a saddle-field source comprises a positively charged annular anode plate set symmetrically between two closed-end cylindrical cathodes (Fig. 9.5(a)). An electron released from rest at an appropriate point within the cylindrical cathode structures can follow an infinitely long oscillatory path, roughly figure-eight-shaped, which always passes through the anode disc aperture. If gas, for example say argon, is

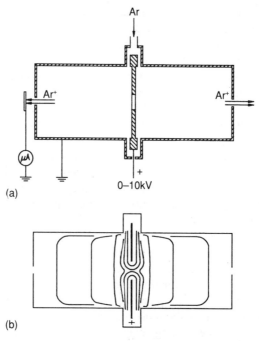

Fig. 9.5 (a) Section through an ion/fast atom source based on electron confinement by an electrostatic saddle field. (b) Equipotential contours of the saddle field.

present within the cylinder, the oscillating electrons cause ionization with high probability, each electron being able to produce a large number of ions and, of course, additional electrons before it is brought to rest at the anode. The electric field existing between the anode and cathode structure accelerates ions away from the anode plane and a beam of ions/fast neutral atoms may be produced, simply by cutting an exit slot or hole in the cathode end surfaces. The equipotentials responsible for the saddle field are sketched in Fig. 9.5(b). The ions, etc., so produced are reasonably monoenergetic and appear with an energy which is a fixed fraction of the applied anode potential. In addition, the beam is only slightly divergent, 4° being typical. Thus, this very simple source requires only a gas supply, (gas pressure in the source ~ 10^{-3}–10^{-4} mbar) and a high voltage power supply; 0–10 kV is convenient. If the gas supply is argon, then argon ions and fast neutral atoms emerge from the source and may be used for sputtering. The substantial fast neutral atom content is particularly useful for sputtering from insulating materials since it does not result in surface charging. Change of the gas supply, to butane say, allows the direct production of diamond-like surface coatings as alluded to previously. Because the device is entirely symmetrical, the ion beam exiting from one end may be allowed to impinge on a monitoring plate which enables a simple measure of device performance through measurement of the ion current. Start-up occurs essentially as soon as the anode potential is applied, the initiating electrons arising from field emission or perhaps cosmic ray impact.

9.4 MOLECULAR BEAM PRODUCTION

One of the most sensitive techniques available for probing the nature of solid surfaces and, in particular, the details of the gas–solid interaction, is that based on molecular beam scattering. A molecular beam provides a fully surface-specific means of studying surface properties, and the interaction spans a range of phenomena from diffraction through inelastic scattering to the irreversible chemisorption described in Chapter 3. The reason for the surface-specific character of the beam–surface interaction is the low kinetic energy of the beam (less than 0.1 eV) which means that the atoms or molecules constituting the beam are physically unable to penetrate the solid and show an extreme sensitivity to the nature of the outermost layer of the solid, be this just the atoms of the solid itself or the gas atoms/molecules which are attached thereto. It turns out also that the scattering cross-section of atoms or molecules adsorbed on a surface is substantially higher than the value for the same atoms/molecules in the gas phase. This makes the beam–surface interaction very sensitive to the presence of adsorbed impurity atoms, a sensitivity which is in excess of that obtainable using alternative surface analytical techniques such as AES (Auger electron spectroscopy) or LEED (low energy electron diffraction).

The central problem in the development of molecular beam technology has been the need to find the beam sources which are both essentially monoenergetic and yet have a high intensity. The requirement for a monoenergetic beam arises from the need to be able to interpret the results of scattering on a par with surface techniques using charged beams of, say, ions or electrons, where the requirement for a monoenergetic beam is readily met. On the other hand, a high beam intensity is required simply to improve detectability. Detection of molecular beams is usually carried out using a modified quadrupole mass spectrometer such as has been described in Chapter 5. It was not until 1951 that the solution to this problem was found with the introduction of the nozzle beam source. Prior to this time, only effusive sources were available whose output was of low intensity and far from monoenergetic.

In an effusive source, gas atoms/molecules merely effuse from an oven or source through an orifice or an array of tubes and then pass through a collimating aperture to produce a molecular beam. In this type of source, the gas pressure driving the source is adjusted to give a Knudsen number in the source greater than unity so that we have free-molecule flow through the source. There is no significant mass transport in the direction of the beam and the velocity (energy) distribution of the beam is necessarily just Maxwellian and characteristic of the oven or orifice temperature. The physical arrangement is shown schematically in Fig. 9.6(a).

In a nozzle source, however, there is net mass transport through the nozzle and this involves the conversion of the total enthalpy into beam translational energy, yielding higher beam energies. A consequence of this mode of operation is that the beam is cooled during expansion through the nozzle and the velocity distribution of the molecules along

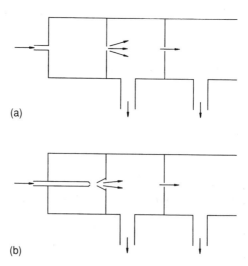

(a)

(b)

Fig. 9.6 Schematic views of (a) an effusive or Knudsen source; (b) a supersonic jet source.

the beam axis, although still approximately Maxwellian, is that appropriate to a temperature much lower than that of the nozzle; it is, in fact,
a Maxwellian velocity distribution superimposed on an average flow
velocity u_s. In other words, rotational, vibrational and some translational degrees of freedom are frozen out and converted into translational motion along the beam axis.

In the nozzle source we have a nozzle or jet aligned with a hollow
truncated cone known as a 'skimmer' which serves to skim off all those
molecules not directed along the beam axis. Figure 9.6(b) shows
schematically the experimental arrangement. In contrast with the
Knudsen source, the gas flow in the nozzle source operates at Knudsen
numbers less than unity as the gas is expanded isentropically through
the nozzle from a high pressure to the vacuum chamber. As the gas
expands into the lower pressure (vacuum) region, lower gas densities
result with increasing distance from the nozzle. The result is that free-
molecule flow eventually takes over from the continuum flow in the
nozzle, as the collision frequency decreases; the beam becomes supersonic in the process and will be characterized by a Mach number M_n
greater than unity. The Mach number here is the ratio of the flow
velocity u_s to the local sonic velocity. Figure 9.7 shows how this
isentropic expansion to Mach numbers greater than unity affects the
velocity distribution, with the velocity distribution narrowing as higher
Mach numbers are attained. Of course, in the case of an effusive or
Knudsen source, $u_s = 0$, and M_n must also be zero. The monoenergetic
quality of the beam is often described by the speed ratio

$$S_R = u_s / \Delta u_s$$

which describes the velocity distribution in the beam (not to be con-
fused with the speed ratio used to describe the performance of turbomolecular pumps). Clearly, this quantity determines the velocity
distribution and hence resolution in any scattering experiment. It is
related to the Knudsen number K_n by the equation

$$S_R = \text{const.} K_n^{-(\gamma-1)/\gamma} \tag{9.1}$$

where γ is the ratio of the principal specific heats and the constant also
depends on γ to some extent.

Equation (9.1) essentially says that a highly monoenergetic beam,
that is a high speed ratio, may be obtained either by increasing the gas
pressure or the orifice diameter. Unfortunately, both of these changes
increase the total gas flow into the system so that pumping capacity

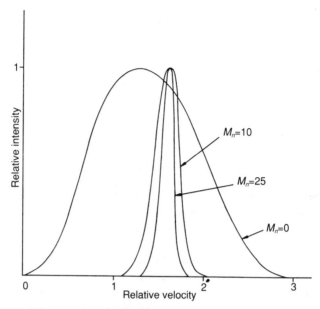

Fig. 9.7 Relative velocity distributions as a function of Mach number for an effusive source $M_n = 0$ and supersonic nozzle sources with increasing Mach number. The relative velocity is just $u(2k_B T/m)^{1/2}$.

ultimately becomes the limitation. The gas flow varies of course with the square of the orifice radius, but only linearly with pressure, as described in Chapter 2, so that usually the speed ratio is increased by working at higher pressures. With a high speed ratio – and speed ratios as high as 200 are possible – it is possible to obtain energy resolution of ~ 200 μeV, which is comparable with that obtained in optical spectroscopy. Indeed, it is possible to demonstrate surface phonon excitation and annihilation during the beam interaction with a surface. A special virtue of the supersonic nozzle source is that as the resolution is increased so is the beam intensity, consequently both resolution and intensity are controlled by the pumping capacity of the system. The pumping requirements of molecular beam systems are rather special for two reasons. The first is a result of the very high gas flow rates encountered, in effect a gigantic permanent leak; the second follows from the fact that the molecular beam systems operate using 'differential pumping'. Reference to Fig. 9.6 will make the principle clearer. It can be seen that in both the effusive source and the nozzle source the beam system is segmented into compartments, each compartment communicating with the next by means of a collimating or defining aperture which ultimately determines the beam's angular divergence. All the molecules which are removed from the beam by the collimating aperture, or the skimmer as well in the case of the nozzle source, must be removed from the vacuum system if they are not to collide with and scatter atoms/molecules already in the beam. This means that each collimating section, and there may be many, must be separately pumped, i.e. each section requires a diffusion pump/turbomolecular pump plus associated backing pumps and pressure measuring equipment. Thus, molecular beam systems may be thought of as pump intensive.

Once an atom/molecule has scattered from a target surface, it must be detected by some means. Generally the experimentalist will need to know the scattering angle and also the energy of the atom/molecule. The former quantity may be obtained by having a system in which the detector, generally a quadrupole mass spectrometer, may be moved around the sample on a goniometer. Energy measurements are most conveniently achieved by means of a time-of-flight measurement using a chopped beam.

9.5 A MOLECULAR BEAM SYSTEM

The basic ideas can be made clearer by consideration of Figure 9.8, which shows in simplified form the high resolution molecular beam system developed by Toennies and others, beginning in 1977. This system operates with a jet nozzle of 5 μm diameter supplied with helium at 200 bar. The first stage is pumped with a 5000 ls^{-1} diffusion pump followed by a further eight stages of differential pumping, with three of these stages between the sample and detector. The pressure in the detector region is of the order of 10^{-12} mbar only, essentially nearly eighteen orders of magnitude pressure difference occurring across the whole system.

A system of this size and complexity does not permit of the detector rotating around the sample, rather the source–target–detector angle is fixed at 90°. The size limitations imposed by these differential pumping requirements are also significant, with the path length between the beam chopper and target being 756 mm and between the target and detector 1015mm, a total path length of 1771 mm. These large flight paths do, however, permit time-of-flight measurements in conjunction

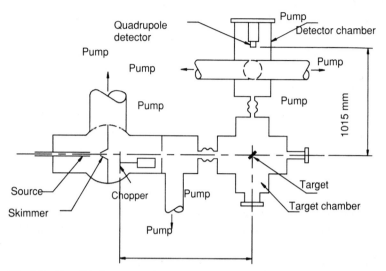

Fig. 9.8 Simplified cross section through a high resolution, molecular beam system based on a supersonic nozzle source. Not all the vacuum pumps are indicated.

with the beam chopper, and the energy resolution of the system is 190 μeV using a liquid nitrogen cooled nozzle.

In operation, the experimental output is a time-of-flight spectrum against signal intensity measured at the quadrupole detector. This spectrum shows up losses and gains in energy by the beam and has been used to demonstrate the existence of interactions with single surface phonons, i.e. the loss of energy from the beam to the surface may result from phonon creation whilst energy gain by the beam is a consequence of phonon annihilation. This is only one aspect of the interaction between high resolution molecular beams and surfaces. If the interaction is an elastic one, then diffraction phenomena may be seen. In this case the scattering is from the electron density contours of the surface rather than from the ion cores of the solid which would be the case for low energy electron diffraction. This is a fine example of a technique originally introduced in the 1930s by Esterman and Stern.

APPENDIX A
Vacuum symbols

Vacuum pumps

 Vacuum pump, general

 Ejector vacuum pump

 Piston vacuum pump

 Diffusion pump

 Diaphragm vacuum pump

 Adsorption pump

 Rotary positive displacement pump

 Getter pump

 Rotary plunger vacuum pump

 Sublimation (evaporation) pump

 Sliding vane rotary vacuum pump

 Sputter-ion pump

 Rotary piston vacuum pump

 Cryopump

 Liquid ring vacuum pump

 Radial flow turbocompressor/ radial flow pump

 Roots vacuum pump

 Axial flow turbocompressor/ axial flow pump

 Turbine vacuum pump, general

 Turbomolecular pump

Connections, tubes and leadthroughs

	Flange connection, general		Crossover of two ducts without connection
	Bolted flange connection		Branch-off point
	Small flange connection		Collection of ducts
	Clamped flange connection		Flexible connection (e.g. bellows, flexible tubing)
	Threaded tube connection		Linear motion leadthrough, flange-mounted
	Ball-and-socket joint		Linear motion leadthrough, without flange
	Spigot-and-socket joint		Leadthrough for transmission of rotary and linear motion
	Connection by taper ground joint		Rotary transmission leadthrough
	Change in the cross section of a duct		Electric current leadthrough
	Intersection of two ducts with connection		

Accessories and valves

Symbol	Description	Symbol	Description
	Condensate trap, general		Shut-off device, general
	Condensate trap with heat exchange (e.g. cooled)		Isolating valve Straight through valve
	Gas filter, general		Right angle valve
	Filtering apparatus, general		Stop-cock
	Baffle, general		Three-way stop-cock
	Cooled baffle		Right angle stop-cock
	Cold trap, general		Gate valve
	Cold trap with coolant reservoir		Butterfly valve
	Sorption trap		Non-return valve
			Safety shut-off valve

Valve actuation, general symbols and gauges

	Manual operation		General symbol for vacuum
	Variable leak valve		Vacuum gauge head, vacuum measurement, general
	Electromagnetic operation		Vacuum gauge control unit
	Hydraulic or pneumatic operation		Recording vacuum gauge (control unit)
	Electric motor operation		Vacuum gauge (control unit) with analogue indication
	Weight operated		Vacuum gauge (control unit) with digital indication
	Vacuum chamber		Measurement of throughput
	Vacuum bell jar		

* The asterisk indicates that the symbol must be used this way up.

APPENDIX B

Brief outline of some vacuum material properties

The values given should be regarded as typical.

Alumel (Al 2%, Ni 95%, Mn 2%, Ag 1%); m.p. 1390° C; density 8.5×10^3 kgm^{-3}; resistivity 28.1×10^{-4} Ωm at 0° C; outgassing rate $10^{-9} - 10^{-7}$ Wm^{-2} after baking. Nickel alloy readily spot welded, usually used as an element of the Chromel–Alumel thermocouple.

Alumina (Al$_2$O$_3$); mol wt 102; m.p. 2020 °C; density 3.7×10^3 kgm^{-3}; resistivity 10^{24} Ωm at 0° C; outgassing rate $10^{-9} - 10^{-6}$ Wm^{-2} after bakeout. The most widely used ceramic in vacuum technology. High mechanical strength, very low gas permeability. Ceramic–metal seals most commonly use alumina combined with Nilo-K (Kovar).

Aluminium At. wt 26.98; m.p. 660° C; density 2.7×10^3 kgm^{-3}; resistivity 2.45×10^{-4} Ωm at 0° C; outgassing rate 10^{-10} Wm^{-2} after bakeout, 10^{-6} Wm^{-2} before bakeout. A ductile metal which is difficult to degas thoroughly owing to its low m.p. Not usually used in pure form for constructional purposes. Vigorously attacked by mercury. Readily evaporated,

Beryllium Copper

low permeability. Very useful for the production of wire vacuum seals if very pure wire (99.99%) is used.

(Be 2%, Co 0.5%, Cu 97.5%); m.p. 1000° C; density 8.2×10^3 kgm^{-3}; resistivity 7×10^{-4} Ωm. Copper alloy, extremely ductile in annealed state but strong and hard after aging. Very good electrical and thermal conductor with very high resistance to corrosion and wear, resulting from a thin surface film of BeO. Excellent for springs, contacts and diaphragms.

Boron Nitride

(BN); mol. wt 24.8. Available in two forms; the material with a graphite structure is used in vacuum work since it has self-lubricating properties, a consequence of its structure. Excellent insulator, with unusually high chemical and thermal stability. May be heated to 1000 °C in vacuum.

Buna N

(Nitrile rubber, copolymer of butadiene and acrylonitrile); outgassing rate 4.7×10^{-2} Wm^{-2}; permeability 4.8×10^{-12} m^{2}s^{-1}. Maximum operating temperature 80° C. Useful as gasket or O-ring material at modest vacua, high resistance to compressive set.

Chromel

(Ni 90%, Cr 10%); m.p. 1420° C; density 8.7×10^3 kgm^{-3}; resistivity 70×10^{-4} Ωm at 0° C; outgassing rate $10^{-9} - 10^{-8}$ Wm^2 after bakeout. Nickel chromium alloy principally used for thermocouple manufacture in conjunction with alumel.

Constantan

(Cu 60%, Ni 40%); m.p. 1210° C; density 8.9×10^3 kgm^{-3}; resistivity 48×10^{-4} Ωm at 0° C; outgassing rate $< 10^{-8}$ Wm^{-2} after baking. Copper–nickel alloy used as a thermocouple material in conjunction with copper. Electrical resistance does not vary with temperature.

Copper (OFHC); at. wt 63.5; m.p. 1083° C; density 8.96×10^3 kgm^{-3}; resistivity 1.6×10^{-4} Ω m; outgassing rate $10^{-7} - 10^{-5}$ Wm^{-2} before bakeout, $10^{-11} - 10^{-8}$ Wm^{-2} after bakeout. Oxygen free, high conductivity copper (OFHC) is the preferred choice for vacuum work, especially for electrodes and gaskets. Excellent thermal and electrical conductivity makes copper an ideal choice for specialized vacuum tube electrodes.

Gold At. wt 196.97; m.p. 1064° C; density 19.3×10^3 kgm^{-3}; resistivity 2.04×10^{-4} Ω m at 0° C; outgassing rate 3×10^{-8} Wm^{-2} after bakeout. Principal use is as O-rings in stainless steel vacuum systems. Other uses are as low reactivity, low secondary emission coefficient coatings for electron tube assemblies. Readily evaporated onto glass and other materials. Attacked by mercury vapour.

Indium At. wt 114.8; m.p. 156.6° C; density 7.31×10^3 kgm^{-3}; resistivity 8.4×10^{-4} Ωm at 0° C. Mainly used for the construction of wire seals where high temperature bakeouts are not required since it has a much lower permeability and outgassing rate than any elastomer.

Iridium At. wt 192.2; m.p. 2410° C; density 22.4×10^3 kgm^{-3}; resistivity 4.9×10^{-4} Ωm at 0° C. Metal of the platinum family. Very hard and brittle. Primarily used for the preparation of filaments for ion gauges, etc. Most corrosion-resistant metal known.

Kalrez® (Perfluoroelastomer); similar in many respects to Viton A®, but possesses superior properties with regard to bakeout. May be baked under compression at 250° C.

Kovar   (Fe 53.7%, Ni 29%, Mo 17%, Mn 0.3%); m.p. 1450° C. Much used for glass–metal seals, particularly for special borosilicate glasses such as Kodial. Magnetic, poor conductor of heat or electricity. Properties similar to Nilo-K.

Mercury   At. wt 200.6; m.p. −38.8° C; density 13.5×10^3 kgm^{-3}; resistivity 94×10^{-4} Ωm at 0° C; vapour pressure 2×10^{-3} mbar at 25° C. Finds widespread use in manometers, McLeod gauges and mercury vapour diffusion pumps. Forms amalgams with many metals. Cannot be used in vacuum systems without liquid nitrogen cold traps owing to very high vapour pressure.

Molybdenum   At. wt 95.94; m.p 2617° C; density 10.2×10^3 kgm^{-3}; resistivity 5.2×10^{-4} Ω m at 0° C ; outgassing rate 10^{-6} Wm^{-2} before bakeout. Hard refractory metal; softer and more ductile than tungsten. Spot-welds to most metals but not itself or tungsten. Like tungsten it oxidizes readily in air. Usually used for gauge filaments or electrode structures.

Molybdenum disulphide   (MoS_2); mol. wt 160.07; m.p. 1185° C; density 4.8×10^3 kgm^{-3}. Black powder with graphite-like structure and lubricating properties, used as a dry, high temperature lubricant for stainless steel screws, etc.

Mumetal   (Ni 77%, Cu 5%, Fe 14%, Mo 4%); m.p. 1440° C; density 8.8×10^3 kgm^{-3}; resistivity 60×10^{-4} Ωm; outgassing rates 10^{-9} – 10^{-7} Wm^{-2} after bakeout. High magnetic permeability, often used to screen low energy electron guns or low energy electron sources from external magnetic fields.

Neoprene   (Chloroelastomer); outgassing rate 40 Wm^{-2}; permeability 0.21×10^{-12} m^{2}s^{-1}.

May be used in low vacuum envir-
onments as a gasket or O-ring material.
Not now widely used.

Nilo-K
(Ni 29%, Co 17%, Fe 54%); m.p.
1450° C; density 8.16×10^3 kgm^{-3}; resistivity 46×10^{-4} Ωm; outgassing rate 10^{-9}
– 10^{-8} Wm^{-2} after bakeout. Used mainly
for glass–metal seals owing to its
wettability by glass and similar expansion coefficient. Magnetic and a poor
conductor of heat or electricity.

Palladium
At. wt. 106.4; m.p. 1552° C; density
12.0×10^3 kgm^{-3}; resistivity $9.93 \times
10^{-4}$ Ω m. Very ductile metal, similar in
many respects to platinum in its properties. Chiefly significant for its ability to
permeate hydrogen selectively at very
high rates and for its very high solubility
for hydrogen; it takes up roughly 900
times its own volume of hydrogen at STP.

Perbunan
(Nitrile rubber, copolymer of butadiene
and acrylonitrile); outgassing rate
4.67 Wm^{-2}; permeability 0.8×10^{-12} m^{2}s^{-1}.
Useful as a gasket or O-ring material in
low vacuum situations. Maximum operating temperature 80° C. Very resistant to
compressive set.

Platinum
At. wt 195.09; m.p. 1769° C; density
21.45×10^3 kgm^{-3}; resistivity $9.81 \times
10^{-4}$ Ωm at 0° C; outgassing rate $2 \times
10^{-8}$ Wm^{-2} after baking. Very ductile
mobile metal, almost totally inert to acid
attack. Often use as a layer between
tungsten or molybdenum surfaces to
permit spotwelding.

PTFE (Teflon®)
(Fluorocarbon); m.p. 330° C; density
2.1 kgm^3; resistivity $> 10^{24}$ Ωm at 0° C;
outgassing rate 10^{-6} Wm^{-2} before
bakeout, 10^{-9} Wm^{-2} after; permeability
2.5×10^{-12} m^{2}s^{-1}. Very inert, slippery,
polymer, widely used for providing self-

lubricating bearings in vacuum systems or as a low-temperature electrical insulator.

Pyrex

(Borosilicate glass); m.p. 555° C; density 2.23×10^3 kgm^{-3}; resistivity 10^{22} Ωm; outgassing rate 10^{-3} Wm^{-2} before bakeout, 10^{-7} Wm^{-2} after bakeout. The basis of all-glass vacuum systems, readily worked by usual glass-blowing techniques.

Quartz

(SiO$_2$); mol. wt 60.08; m.p. 1610° C; density 2.6×10^3 kgm^{-3}; resistivity 10^{24} Ωm; outgassing rate before bakeout $10^{-4} - 10^{-2}$ Wm^{-2}, $10^{-8} - 10^{-6}$ Wm^{-2} after bakeout. Main use in optics where high uv transmission is required or as insulating elements in electron guns.

Rhenium

At. wt 186.22; m.p. 3180; density 21×10^3 kgm^{-3}; resistivity 18.6×10^{-4} Ωm at 0° C. Ductile metal commonly used as a replacement for tungsten as a thermionic emitter filament. Unaffected by heating in air and consequently does not readily burn out. Excellent as a corrosion-free surface finish if either evaporated or electroplated.

Sapphire

(Al$_2$O$_3$); at. wt 102; m.p. 2040° C; density 3.98×10^3 kgm^{-3}; resistivity $> 10^{24}$ Ωm; outgassing rate $10^{-9} - 10^{-7}$ Wm^{-2}. Pure crystalline form of alumina. Excellent for use as wide transmission band optical windows or as insulator. Available as precision ground balls.

Silver

At. wt 107.87; m.p. 960.8° C; density 10.5×10^3 kgm^{-3}; resistivity 1.51×10^{-4} Ωm; outgassing rate before bakeout 10^{-6} Wm^{-2}, 10^{-9} Wm^{-2} after bakeout. Excellent conductor of heat and electricity. Useful brazing material in the pure form or as the silver/copper eutectic (Ag 72%, Cu 28%) for brazing copper. Readily permeable to oxygen.

Stainless steel (Fe 71%, Cr 18%, Ni 10%, Mn/Si/C 1%); m.p. 1430° C; resistivity 72×10^{-4} Ωm; outgassing rate before bakeout 10^{-7} – 10^{-5} Wm^{-2}, after bakeout 10^{-11} – 10^{-8} Wm^{-2}. Primary construction material for vacuum systems, particularly UHV systems, owing to its low permeability to hydrogen, resistance to corrosion and ease of bakeout.

Tantalum At. wt 180.95; m.p. 2996° C; density 16.6×10^3 kgm^{-3}; resistivity 12.6×10^{-4} Ω m. Ductile, malleable, highly refractory metal. Strength comparable with that of mild steel. Finds use in manufacture of electrode assemblies where high temperatures are anticipated. Useful as an intermediary metal in the spot welding of tungsten to tungsten or molybdenum to molybdenum. Very resistant to corrosion or chemical attack.

Titanium At. wt 47.9; m.p. 1675° C; density 4.5×10^3 kgm^{-3}; resistivity 50×10^{-4} Ωm; outgassing rate $< 10^{-9}$ Wm^{-2} after bakeout. Lustrous white metal, easily fabricated, resistant to corrosion. Excellent gettering material.

Tungsten At. wt 183.85; m.p. 3410° C; density 19.3×10^3 kgm^{-3}; resistivity 4.9×10^{-4} Ωm; outgassing rate after bakeout 5×10^{-8} Wm^{-2}. Most refractory metal known, oxidizes readily in air. Commonly used as a thermionic filament although it becomes brittle after firing in vacuum. Increasingly being superseded by rhenium or iridium in this function.

Viton A® (Fluoroelastomer); density 1.8×10^3 kgm^{-3}; resistivity 2×10^{21} Ωm; vapour pressure 5×10^{-8} mbar at 150° C. Should not be baked under compression at temperatures $> 150°$ C. Takes compressive set. Degrades slightly at 192° C.

Zirconium   At. wt 91.22; m.p. 1852° C; density 6.5×10^3 kgm^{-3}; resistivity 40×10^{-4} Ωm. Grey-white metal. Properties very similar to titanium, particularly its gettering ability. Resistant to corrosion.

APPENDIX C

Some vacuum equipment manufacturers and suppliers

Alcatel Vacuum Engineering
The Business Centre Building, No.2 The Vansittart Estate, Windsor, Berks, SL4 1SE, UK.
General vacuum components; pump manufacturer.

Balzers High Vacuum Ltd
Northbridge Road, Berkhamstead, Herts, UK.
General vacuum components; pump manufacturer.

Caburn MDC Ltd
1, Castle Ditch Lane, Lewes, East Sussex, BN7 1YJ, UK.
General vacuum components; specialist vacuum system manufacturer.

CVT Ltd
4, Carters Lane, Kiln Farm, Milton Keynes, MK11 3ER, UK.
General vacuum components; specialist vacuum system manufacturer.

Edwards High Vacuum International
Manor Royal, Crawley, West Sussex, RH10 2LW, UK.
General vacuum components; pump manufacturer; specialist vacuum system manufacturer.

Granville Philips
5675 Arapahoe Avenue, Boulder, Colorado, 80303-1398, USA.

General vacuum components, pump manufacturer.

Hiden Analytical Ltd   10, Greys Court, Kingsland, Grainge, Warrington, WA1 4RW, UK.

Mass spectrometers.

Leybold AG   Bonner Strasse 498, D5000, Cologne 51, West Germany.

General vacuum components; pump manufacturer; specialist vacuum equipment manufacturer.

Vacuum Generators Ltd   Maunsell Road, Hastings, East Sussex, TN38 9NN, UK.

General vacuum components; specialist vacuum system manufacturer; mass spectrometers.

Varian Vacuum Products   121, Hartwell Avenue, Lexington, Massachusetts, 02173, USA.

General vacuum components, pump manufacturer.

Veeco Instruments Inc.   Terminal Drive, Plainview, New York 11803, USA.

General vacuum components, leak detection systems, evaporation systems.

Answers to problems

1 Aspects of kinetic theory

1.1 0.53; 1.8×10^4 s

1.2 0.055 Pa

1.3 3.85 μm contraction; if pressure doubled, contraction doubles to 7.70 μm

1.4 No change in λ, gas density unaltered

2 Flow of gases through tubes and orifices

2.1 5×10^{-4} mbar

2.2 2727 ls^{-1}

2.3 679.4 ls^{-1}

2.4 90.2 ls^{-1}

3 Physisorption, chemisorption and other surface effects

3.1 287 picogram

3.2 2.49×10^{-12} Pa; 2.5×10^{-10} Pa

3.3 2.5×10^3 ls^{-1}; 2.5×10^6 ls^{-1}

4 Vacuum pumps, the physical principles

4.1 1.65×10^3

4.2 300 lmin^{-1}

4.3 (a) 3710 ls^{-1}; (b) 13883 ls^{-1}

5 Pressure measurement

5.1 6.7×10^{-3} mbar

5.2 7.5×10^{-11} mbar

5.3 0.005 ls^{-1}

6 Vacuum systems design

6.1 $S_p' = 1000$ ls^{-1}; $F_p = 6000$ ls^{-1}; $S_p = 1200$ ls^{-1}; $\tau = 0.1$s

6.2 $S_f = 576$ lmin^{-1}; $F_f = 48$ ls^{-1}; $t = 2.4$ min

6.3 9.9 s

8 Leak detection

8.1 Viscous flow leak, 38% pressure drop

Further Reading

Kinetic theory and the basis of vacuum technology has been the subject of many books over the years. The following books offer useful insights, but must not be considered an exhaustive list.

Kennard, E. H. (1938) *Kinetic Theory of Gases*, **McGraw-Hill, New York and London.**

This is an extremely detailed and comprehensive view of the kinetic theory of gases which, despite its age, is still very useful.

Loeb, L. B. (1934) *The Kinetic Theory of Gases*, **Dover Edition (1961), New York.**

This is another very detailed overview of kinetic theory. Although it predates Kennard's volume and Kennard admits an indebtedness to Loeb, it still manages to bring a better physical understanding, for example, to many of the more complex aspects of gas flow. It is only available now as a Dover edition.

Guénault, A. M. (1988) *Statistical Physics*, **Routledge, London.**

This book provides a clear and compact introduction to the Maxwell distribution of velocities.

Dushman, S. (1949) *Scientific Foundations of Vacuum Technique*, **2nd edn. (1961), John Wiley & Sons, New York.**

This book was a classic in its field when it was first published. The 2nd edition of 1961 contains much additional material from members of the research staff of the General Electric Research Laboratory, USA. It is somewhat dated now, but still useful and very wide ranging.

O'Hanlon, J. F. (1980) *A User's Guide to Vacuum Technology*, **John Wiley & Sons, New York.**

As the title suggests this book is orientated towards the vacuum system user. It covers a wide range of material with very good depth of treatment.

Dennis, N. T. M. and Heppell, T. A. (1968) *Vacuum System Design*, **Chapman & Hall, London.**

This is a comprehensive examination of vacuum system technology more from the point of view of someone faced with designing a vacuum system.

Woodruff, D. P. and Delchar, T. A. (1988) *Modern Techniques of Surface Science*, **Cambridge University Press, Cambridge.**

For those readers interested in discovering more of the details of the processes which operate at surfaces and which often dictate vacuum system behaviour.

Redhead, P. A., Hobson, J. P. and Kornelson, E. V. (1968) *The Physical Basis of Ultrahigh Vacuum*, **Chapman & Hall, London.**

This is a very comprehensive view of the problems associated with the attainment and measurement of UHV by some of the pioneers in the field.

Chapman, S. and Cowling T. G. (1991) *The Mathematical Theory of Non Uniform Gases*, **3rd edn, Cambridge University Press, Cambridge.**

This book represents a detailed mathematical account of the kinetic theory of viscosity, thermal conduction and diffusion in gases.

Leck, J. H. (1989) *Total and Partial Pressure Measurement in Vacuum Systems*, **Blackie, Glasgow and London.**

A useful survey of pressure measurement.

Index

Absorbate 48
Absorbent 47
Absorption 47
Accommodation coefficient 16
Activated alumina 65
Activated charcoal 65, 104
Activation energy 57
Active alloy seals 192
Adsorbate 48
Adsorbent 47
Adsorption 47
Adsorption isostere 50
Adsorption isotherm 50, 55
Aluminium 61, 75, 176, 198
Argon instability 103
Average speed 2
Avogadro's law 3

Backstreaming 166, 168, 169
Baffle 166, 194
Bakeable metal valves 188
Bakeout 76, 169, 171
Ball valves 187, 188
Ballast tank 216
Bellows seals 182
Bond strengths 68
Boyle's law 3, 156
Brunauer, Emett and Teller
    (BET) isotherm 54

Cathode memory 103
Ceramic-to-metal seals
    192, 193
Ceramics 194
Charge exchange 70
Charles' law 3
Chemisorption 48, 56, 58, 78
    associative 57
    dissociative 57, 58
Clausing coefficient 33, 36, 42
Clausing factor 44
Coinage metals 58
Cold cathode discharge 71
Cold trap 166, 194
Compression ratio 84, 85, 91, 92
Condensation coefficient 59
Conductance, see Flow
    conductance and impedance
Conflat flange 176
Convectron gauge 123
Conversion constant 140
Copper 58, 61, 198
Critical back pressure 96
Critical backing pressure 96
Cryopump capacity 108

Decker valve 185
Desorption 48, 76
Diamond 223
Differential pumping 228

Diffstak pump   220
Diffusion   73
Diffusion coefficient   21, 22, 74
Diffusion of gases   21
Diffusion seals   192
Dissociative chemisorption   57, 58
Dushmann's coefficient   36

Effusion   33
Effusive source   226, 227
Electrical clean up   69
Electrical feedthroughs   192
Electroflux refining   198
Electron beam heating   220
Electron emission   133
Electron stimulated desorption   72
End correction, tube   37
End effect, tube, *see* orifice conductance
Energetics of physisorption   49
Entrance transition   31
Entrapment   78
Equilibrium pressure   167
Extreme ultrahigh vacuum   140
   see also Vacuum spectrum

Faraday dark space   69, 70
Fast neutral atoms   223
Fast neutral molecule   71
Fernico   191
Fick's law   73
Field emission   137, 138
Field-induced adsorption   69
Flow conductance and impedance   28
Flow conductance
   tubes in parallel   29
   tubes in series   29
Flow in transition range   38
Flow transition region   41

Flow rate 28
Fomblin   85
Free molecule conductance, tubes in series   42
Freundlich isotherm   50

Gas admission   189
Gas calibration factor   142, 209
Gases, kinetic theory of   1
Gases, non-equilibrium properties of   4
Gauge pumping action   133
Gettering   61
Getters   61
Glass-to-metal seals   190, 194
Gold   58, 61, 176, 198

Heat conduction, free-molecule   19
Heat of adsorption   49, 56
Heat transfer   123
Henry's law   51, 53
Hertz–Knudsen equation   23, 52, 106
Housekeeper seal   191, 193
Hydrogen   75
Hyperbolic isotherm   53, 54

Ideal gas laws   2
Impedance to flow   29
Inchworm motor   184
Indium   176, 198
Intrinsic speed   44
Ion current   128, 129, 132, 134, 135, 139
Ion stimulated desorption   72
Ion trap   151
Ionization cross section   73
Ionization potential   128
Ionization probability   129
ISO–F clamp flange   179
ISO–K clamp flange   179

Kalrez   170, 177
Klein flange   178
Knudsen flow   27, 38, 40
Knudsen number   26, 27
Knudsen source, *see* Effusive
      source
Kovar   191

Lafferty gauge, *see* Pressure
      gauge, hot cathode
      magnetron
Langmuir isotherm   53
Langmuir model   52
Lanthanum hexaboride   133, 141
Leak detection
   by partial pressure analysis
      211
   'bombing'   218
   dynamic   201, 208
   'sniffing'   217
   'spraying'   218
   static   210
Leak test system, portable   213,
      214
Leak unit   204
Leak
   minimum detectable   201,
      202, 211
   real   200
   virtual   200
Linde sieves 4A, 5A, 13X   67,
      104
Liquid helium   107
Long tube   31, 32

Mach number   29, 227
Magnetic coupling   183, 185
Magnetic levitation   94, 169
Magnetron, conical   222
Manometer
   capacitive   116

Rayleigh differential   113
   U-tube   113
Mass spectrometer
   magnetic deflection   144
   monopole   148
   omegatron   146
   quadrupole   148, 149
Mathieu equation   149
Maxwell's distribution of
      velocities   2
Mean free path   5
Metal valve   187
Metals for vacuum use   196
Microbar   4
Millimetre Hg   4
Minimum backing pressure
      162
Minimum backing pump speed
      162
Molecular beam scattering   225
Molecular beam system   229
Molecular flow conductance   32,
      39
Molecular flow   27, 32, 33, 37,
      160, 202
Molecular flow through short
      tubes   33
Molecular sieves   65, 66, 104
Molecule collision frequency, *see*
      Hertz–Knudsen equation
Molybdenum   198
Molybdenum disulphide   176
Momentum flow   8
Most probable speed   2
Multilayer adsorption   54

Negative glow   70
Nickel   61, 189
Nottingham   131
Nozzle beam source   226, 227,
      228

O-ring seals 181
O-rings 180
Observed pumping speed 45
Occlusion 47
OFHC copper 176
Optical windows 192
Orifice conductance 34, 35
Outgassing 76

Palladium 189
Parabolic adsorption isotherm 50
Parallel connection 29
Partial pressure gauge 143
Pascal 4
Penning discharge 101
Perbunan 178
Permeability 74, 75
Permeation 74, 75, 156, 168, 170, 189
Persorption 66
Physisorption 48, 50, 51, 78
Piezoelectric coupling 183
Pipe couplings 175
Plasma excited sputtering 221
Poiseuille's equation 29, 30
Potassium 62
Pressure gauge
  absolute 112
  Bayard–Alpert 131
  cold cathode ionization 136
  diaphragm 116
  high pressure ion gauge 135
  hot cathode ionization 127, 130, 131
  hot cathode magnetron 139
  inverted magnetron 137
  ionization 127, 128
  Knudsen 119
  magnetron ionization 137
  Mcleod 114

measuring limits 111
modulated Bayard–Alpert 133
partial 110
Penning 136
Pirani 122
Redhead magnetron 112, 137
Schulz–Phelps 135
spinning rotor gas friction 120
thermal conductivity 127
thermcouple 125
thermistor 126
total 110
Vacustat 115
Wallace and Tiernan 116
Pressure limit 115, 120, 123, 127, 131, 132, 135, 138, 140, 141, 150
Probe gas 204, 208, 210
Pump-down time 158, 161, 162
Pump oil 85
Pump throughput 44
Pump
  backing 79, 163
  cryopump 79, 105, 166
  diffusion 79, 94, 166
  diode/triode 103
  fore 86, 162
  gas ballast 84
  getter 100
  getter-ion 79, 100, 169
  hook and claws 78, 86
  molecular drag 88
  multistage 83, 84
  multistage vapour diffusion 97
  oil-sealed rotary 78
  Roots 79, 87, 165
  rotary oil 80
  rotating eccentric piston 82, 84

Pump (*cont.*)
  self-fractionating 98
  single-vane rotary 82
  sorption 78, 104, 165
  sputter-ion 102
  sublimation 103
  triple-vane rotary 82
  turbomolecular 79, 9, 166, 169
  vapour ejector 99
Pumping fluid 98, 166
Pumping speed 45, 81, 83, 93, 97, 99, 103, 106, 155, 162
  effective 157
  intrinsic 154
Pumps, backing properties of 165
Pumps, high vacuum, properties of 172

Quadrupole, three-dimensional 151
Quartz 189

Rate of adsorption 52
Rate-limiting processes 76
Reflection coefficient 11
Residual gas analyser 143
Residual gas spectrum 145
Residual pressure 169
Resolving power 143
Reynolds number 31, 32
Rhenium 198
Rhodium 61
Root mean square speed 2
Rotational motion 183

Saddle-field confinement 223
Saddle-field ion/fast neutral source 223
Scattering cross section 225

Sensitivity 139, 140, 144
Series connection 29
Silver 58, 61, 189
Sintered metal seals 192
Skimmer 229
Slip coefficient 9, 12, 15, 38
Slip distance 9
Sodium 62
Sorbate 48
Sorbent 47
Sorbents and molecular sieves 65
Sorptive ability 68
Sorptive capacity 48
Speed of a pump 44
Speed of exhaust 157
Speed of sound 95
Speed ratio 92, 227
Sputtering 63, 64, 100, 102, 221, 223
Sputtering coefficient 65
Sputtering yield 64
Stainless steel 76, 197
Standard atmosphere 4
Stepper motors 185
Sticking coefficient 59, 60, 61, 100
Sticking probability 59
System time constant 158, 207

Temperature jump distance 14, 15, 18, 19
Tesla coil 202, 207
Thermal conductivity 17, 124
  at low pressures 14
  at ordinary pressures 13
Thermal transpiration 23, 24
Thermomolecular flow 23
Throughput 28
Titanium 61, 63, 65, 100, 102
Transition flow 38

Transition metals   58
Transport of energy   5
Transport of mass   5, 21
Transport of momentum   5
Trap, activated alumina   164, 195
Trap, molecular sieve   164
Tube conductance   159
Tungsten   198

Ultimate pressure   154, 157
Ultrahigh vacuum   75, 13, 167,
    169, 171
Ultrahigh vacuum system   172,
    173
Units of pressure   4

Vacuum-based coating systems
    220
Vacuum seals   170
    dynamic   181
Vacuum spectrum   110
Vacuum system design   161
Vacuum system   153
    bell jar   214
Vacuum valves   186
Van der Waals forces   48, 68

Viscosity   9
    coefficient of   7, 12, 22
    free molecule   12
Viscosity of gases   9
    at ordinary pressures   7
    at very low pressures   11
Viscous conductance   38
Viscous flow   26, 29, 31, 33,
    160, 202
Viton   170, 177, 180
Volume polarization   49

Weld preparation   197
Wilson seals   181
    see also Vacuum seals
Wire seal   177
    see also Vacuum seals

X-ray photocurrent   132, 134,
    140

Yarwood's approximation   33

Zeolites   65
    see also Molecular sieves
Zirconium   100